INSTALL YOUR OWN

SOLAR
PANELS

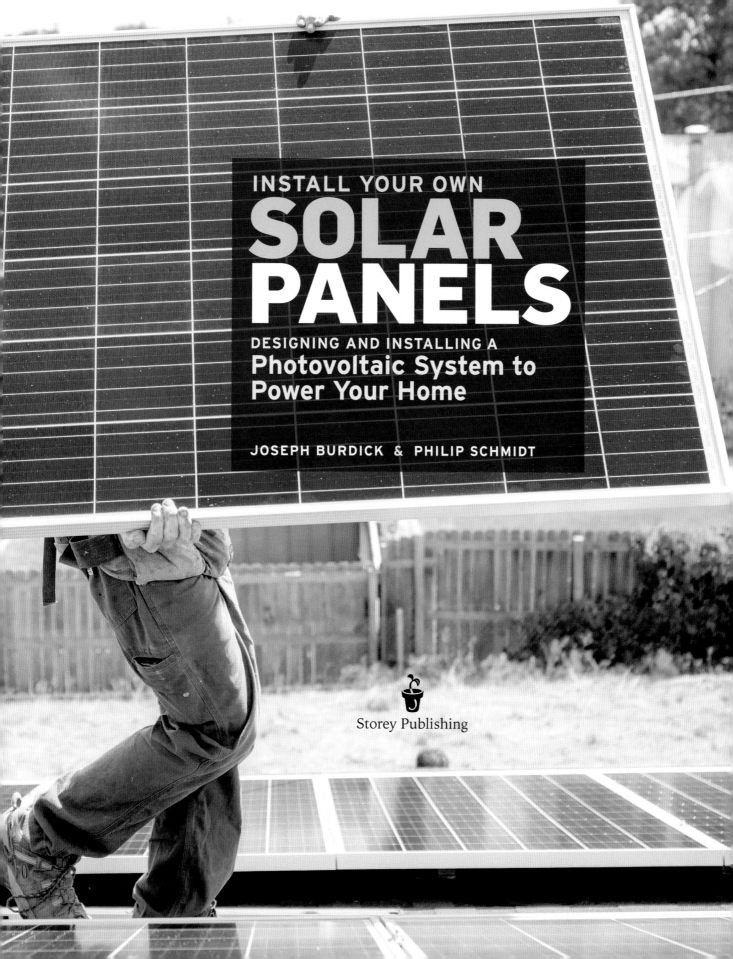

INSTALL YOUR OWN
SOLAR
PANELS

DESIGNING AND INSTALLING A
Photovoltaic System to
Power Your Home

JOSEPH BURDICK & PHILIP SCHMIDT

Storey Publishing

The mission of Storey Publishing is to serve our customers by publishing practical information that encourages personal independence in harmony with the environment.

EDITED BY Deborah Burns and Hannah Fries
ART DIRECTION AND BOOK DESIGN BY Carolyn Eckert
TEXT PRODUCTION BY Liseann Karandisecky
INDEXED BY Nancy D. Wood

COVER PHOTOGRAPHY BY © Heshphoto, inc. except © suze/photocase.com, front (bottom) and back (right)

INTERIOR PHOTOGRAPHY BY © Heshphoto, inc.

ADDITIONAL PHOTOGRAPHY BY © Andrew Wickham/SnapNRack, 123 (top left), 138 (right); © anmbph/iStockphoto.com, 101; © Arina P. Habich/Shutterstock, 47 (bottom left); © Benjamin Schneider, 168; © Bird Barrier America, Inc., 41; © Burdick Technologies Unlimited, 15, 44, 47 (middle), 54, 116, 124-125, 143; © c12/iStockphoto.com, 47 (top); © DPW Solar, 47 (bottom right); © Frase Electric, LLC, 52 (top); © Fronius USA, LLC, 50 (top), 181; © IronRidge, 40 (left); © Joe_Potato/iStockphoto.com, 131; © Mark Miller Photos/Getty Images, 127; Mars Vilaubi, 27, 86; © Midnite Solar, Inc., 155; © Niels Poulsen std/Alamy Stock Photo, 39 (top); © Orion Solar Racking, 45; © panic_attack/iStockphoto.com, 38 (bottom); © Quick Mount PV, 42, 138 (top left); © RSTC Enterprises, 138 (bottom left); © S:FLEX, Inc., 17; © scanrail/iStockphoto.com, 39 (bottom); © SolarEdge, 52 (bottom); © Space Images/Getty Images, 137; © Trojan Battery, Co., LLC, 158; WhistlingBird/Wikimedia Commons, 38 (top); © Wiley/BURNDY, 43 (bottom); © Zomeworks Corporation, 28

ILLUSTRATIONS BY © Michael Gellatly

This publication is intended to provide educational information for the reader on the covered subject. Be sure to read all explanations and advice in this book and in your product literature thoroughly *before* beginning to install your own solar panels. The authors and publisher advise that anyone who is not a trained electrician get professional assistance before connecting to your home's electrical system, and follow all applicable state and local building codes.

Storey Publishing
210 MASS MoCA Way
North Adams, MA 01247
storey.com

Printed in China by R.R. Donnelley
10 9 8 7 6 5 4

Library of Congress Cataloging-in-Publication Data

Names: Burdick, Joesph, 1968- author. | Schmidt, Philip, author.
Title: Install your own solar panels : designing and installing a photovoltaic system to power your home / Joesph Burdick and Philip Schmidt.
Description: North Adams, MA : Storey Publishing, [2017] | Includes bibliographical references and index.
Identifiers: LCCN 2016059736 (print) | LCCN 2017023687 (ebook) | ISBN 9781612128269 (ebook) | ISBN 9781612128252 (pbk. : alk. paper)
Subjects: LCSH: Photovoltaic power system—Design and construction. | Photovoltaic power system—Installation.
Classification: LCC TK1087 (ebook) | LCC TK1087 .B87 2017 (print) | DDC 696—dc23
LC record available at https://lccn.loc.gov/2016059736e

For Daniel, Billie, and Lisa Burdick,
for living exemplary lives that give me continued inspiration.
— JB

For Diana,
It shines on us, too, and helps us all grow.
Infinity Googolplex
— PS

CONTENTS

INTRODUCTION

If you're interested in this book, surely you already know that solar electricity is good for the environment, national security, and the air we breathe, not to mention your electricity bill. And that it's one of the best ways to reduce your household's contribution to global warming. You've also probably heard that going solar can actually be cheaper than paying for utility power, and you might wonder whether this claim is true. Well, in most cases, it is true. It just takes time for the incremental savings to overtake the initial investment (after that, the solar power is free). If you install the solar system yourself, you can hit this tipping point a lot sooner – in some cases, in half the time.

That brings us to the next big question: Can you really install your own solar panels? Again, the answer is yes. If you can drive lag bolts and assemble prefabricated parts, and if you're willing to spend a day or two on your roof (or not, if you're mounting your panels on the ground), you can install your own solar system. You don't have to know how to hook up the solar panels to your household electricity or the utility grid. You'll hire an electrician for the house hookup, and the utility company will take care of the rest, usually for free.

For a completely off-grid system, the utility company isn't involved at all.

Perhaps disappointingly, this job isn't even a good excuse to buy new power tools, since the only one you need is a good drill.

So, if this is such a doable project, why do most people use professional installers? For starters, a lot of people have good reasons to hire out virtually everything, from oil changes to grocery shopping. (That's probably not you, but even if it is, this book can help you plan for a solar installation and find a good local installer.) Solar professionals handle more than the installation. They design the system, they apply for rebates and credits, they order all the necessary parts, and they obtain the permits and pass all the inspections. But the fact is, you can do all of these things yourself, provided you have a helpful adviser (this book, for example) and you are willing to follow the rules of the local building authority (that's where you'll get those permits).

Solar installations are getting easier all the time, and you might be surprised at how much do-it-yourself (DIY) help is available. Two good examples are PVWatts and the Database of

State Incentives for Renewables & Efficiency (DSIRE). PVWatts is an online calculator that helps you size a solar-electric system based on the location and position of your house and the angle of your roof. Solar pros use the same simple tool, but it's free for everyone. DSIRE offers an up-to-date, comprehensive listing of renewable energy rebates, tax breaks, and other financial incentives available in any area of the United States. And it's also free and easy to use.

Those two resources alone help answer the two most common questions homeowners have about solar electricity: *How big of a system do I need?* and *How much will it cost?* Other resources include solar equipment suppliers that cater to DIYers and offer purchasing and technical support, as well as consumer-friendly industry sources like *Home Power* magazine and the online community Build It Solar. And there's no law that says DIYers can't hire a solar professional for help with specific aspects of their project, such as creating design specifications, choosing equipment, or preparing permit documents.

What's been missing from the wealth of information out there is a single expert voice — a guiding light, if you will — that pulls it all together and walks you through the project from start to finish. This book is your guiding light. It tells you what you need and *shows* you how to install it. Every major aspect of the process is covered, from site assessment and system design to permits, equipment, and installation.

We should also say up front that this book is not for cutting corners. We don't want you to install your system without a permit or without hiring an electrician to make the final hookups. (Even professional solar installers use electricians for this stuff.) The permit process can be a pain, yes, but it's there to ensure that your system is safe, not just for you but also for emergency responders who might need to work around your mini power plant. When you work with the local building department you also learn about critical design factors, such as wind and snow loads, that are specific to your area.

Most likely, you're looking to install a standard grid-tied solar-electric system or an off-grid system. We cover both. We also cover the two most common installation scenarios: rooftop and ground-mount. The information is practical, visual, and as nontechnical as possible, teaching you just what you need to know to get the job done. (If you're the type to geek out on things like battery chemistry or multijunction solar cells, you'll probably want some additional sources.) To keep this big job in perspective, each chapter starts with a simple goal that summarizes what you'll accomplish at that stage in the process.

Of course, your ultimate goal is to power your house with the sun. And the sun? Well, it's ready whenever you are.

CAN YOU REALLY INSTALL YOUR OWN SOLAR PANELS? . . .

THE ANSWER IS YES.

1
The Basics

GOAL

Learn the basic parts of solar power systems and the installation process, and confirm that DIY installation is right for you

WE'LL BEGIN YOUR JOURNEY to electrical enlightenment with a quick overview of what goes into a home solar power system and a look at the specific components of the three main types of systems. Then we'll take a brisk walk through the installation process from start to finish. And that's pretty much all there is for the basics lesson, because then it's time to get to work.

Your first task is to make sure that a do-it-yourself (DIY) installation is not only desirable but also legal in your area. (It's not allowed every-where.) If you give yourself the green light, great. If not, you can use this book to learn the essentials of going solar and gain confidence for choosing a good local professional solar installer and getting what you want. The next task is for everyone: determining how much electricity you use and thinking about where your solar system is most likely to go.

Anatomy of a Solar-Electric System

The science of turning sunlight directly into electricity is known as **photovoltaics (PV)**, referring to **photons** of light and **volts** of electricity. Here's your 10-second lesson on how PV works: Solar panels, properly called PV **modules** (see You Say "Panels"; We Say "Modules," page 10), contain solar cells, which are most commonly made of layers of silicon, a semiconductor material made from sand (also the namesake of Silicon Valley). When photons of light enter a solar cell, they get absorbed and excite electrons in the silicon layers, causing them to move and, ultimately, flow continuously through a circuit of wiring that feeds into the PV system.

Harnessing this electron flow is what gives you electrical power.

The electricity produced by PV modules (and used by all batteries) is **direct current (DC)**, in which all of the electrons move in one direction only. Your home's electrical system and most appliances use **alternating current (AC)** power, in which the electrons move back and forth, alternating direction about 60 times per second. Therefore, PV systems include one or more inverters that convert the DC solar-generated electricity to usable AC power for your home (and, with grid-tied systems, for selling back to the utility grid).

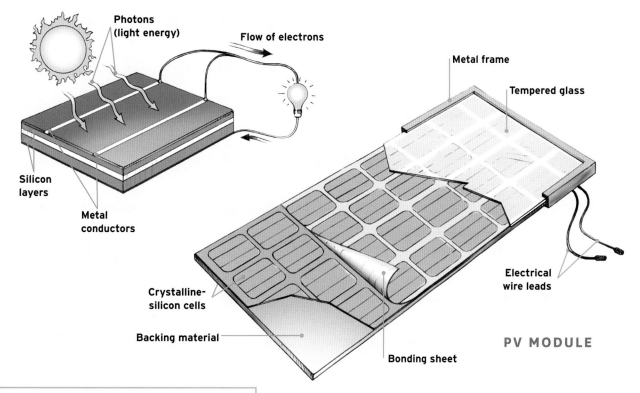

Photons
(light energy)

Flow of electrons

Metal frame

Tempered glass

Silicon
layers

Metal
conductors

Crystalline-
silicon cells

Electrical
wire leads

Backing material

Bonding sheet

PV MODULE

MODULE MAGIC

A standard PV module is little more than a metal frame surrounding a sandwich made of a rigid backing material, thin layers of silicon solar cells, a transparent bonding sheet, and a tempered-glass top. The cells are where the magic happens. Each cell typically measures about 4 to 6 inches square, and a full-size module usually contains 60 or 72 interconnected cells laid out in a grid pattern. The cells are most commonly made of crystalline silicon (c-Si), a semiconductor material capable of producing about 0.5 volt of electricity. (The amount of current, or amperage, produced by the solar cell depends on its size.) When you wire cells together in a series, the voltage adds up so that a 60-cell module produces about 30 volts, and a 72-cell module about 36 volts.

Solar cells are assembled into PV modules for two very good reasons:

1
You need a lot of cells wired together to make a useful amount of electricity produced at a relatively high voltage.

2
Crystalline silicon cells are fragile. The module structure and glass top protect the little magic-makers from things like rain, snow, wind, hail, bird doo-doo, and the long-term effects of temperature and moisture changes. Yes, the modules can break if you drop them the wrong way, but otherwise they're designed to withstand outdoor exposure for several decades.

All home PV systems start with a collection of solar-electric modules, called the PV **array**. The array can be installed on a roof or on the ground. The modules in an array are usually wired together in groups, each called a **series-string**. The series-strings are joined near the array at a combiner box or other device, and wiring from the box brings the power to the rest of the system components at the ground level. The first component that these supply lines connect to depends on the type of system. The following pages give you a snapshot of the three main systems. We'll cover system types and hardware in greater detail in chapters 3 and 4.

Grid-Tied System

A grid-tied system is by far the most common type of residential PV system, as well as the simplest and least expensive. It connects to the electric utility grid and uses the grid for both "storage" and backup. When the array creates more power than the house uses, the excess power is fed back onto the grid – turning the utility meter backward – and you get credited for it. When the house needs more than the solar array provides, the house automatically pulls power from the grid.

Advantages of grid-tied systems include simplicity, low cost, and low maintenance, making them the obvious choice for homeowners who are already using utility power, which is most homeowners. But the grid is also the main disadvantage: when it goes down, so does the PV system. This automatic shutdown function, called self-islanding, is required by utilities for grid hookup for the safety of utility personnel working on the power lines.

Grid-tied systems can use one or more **string inverters**, which convert power from DC to AC for a group of modules at once, or **microinverters**, which convert power from DC

to AC at each individual module or a pair of modules. A third option is to add **DC optimizers** to a string inverter system. DC optimizers (see page 50) add some performance optimization and monitoring features offered by microinverters, but they do not convert DC to AC at the module.

Off-Grid System

The ultimate in self-sufficiency, off-grid systems have no connection to the utility grid and are therefore the best choice for homes far from utility lines. They include a bank of batteries for storing solar-generated power during the day and feeding the house with power at night. These systems also may get additional backup power from a fuel-powered (usually gas, diesel, or propane) generator, which should be installed by an electrician. All solar electricity goes through the batteries; it does not power the house directly from the array. The batteries are charged by DC power from the array and are monitored and controlled by a device called a charge controller. Battery power is converted to AC (through a DC–AC inverter) before supplying the house.

Grid-Tied System with Battery Backup

A grid-tied setup can be combined with battery backup such that solar power charges the batteries and backfeeds the grid when there's an excess. When the house needs more power than the solar array produces, it can pull from the grid or the batteries. When the grid goes down, the batteries supply power to a **critical loads subpanel**, which serves a few household circuits. This enables you to keep important things like the fridge, lighting, computers, and perhaps a gas furnace running during power outages. The batteries typically do not power the entire house, as this would require a larger, more expensive battery bank.

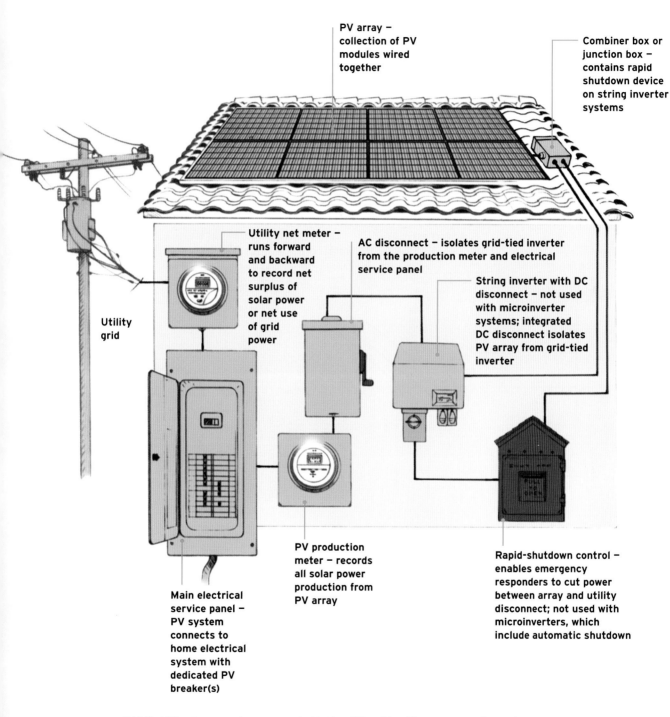

PV array – collection of PV modules wired together

Combiner box or junction box – contains rapid shutdown device on string inverter systems

Utility net meter – runs forward and backward to record net surplus of solar power or net use of grid power

AC disconnect – isolates grid-tied inverter from the production meter and electrical service panel

String inverter with DC disconnect – not used with microinverter systems; integrated DC disconnect isolates PV array from grid-tied inverter

Utility grid

Main electrical service panel – PV system connects to home electrical system with dedicated PV breaker(s)

PV production meter – records all solar power production from PV array

Rapid-shutdown control – enables emergency responders to cut power between array and utility disconnect; not used with microinverters, which include automatic shutdown

Grid-tied PV systems are always connected to the utility grid and have no means for storing energy on-site. A string inverter system is shown here. A microinverter system would have multiple microinverters installed under the array (and no string inverter) and would not have a separate control system for rapid shutdown.

ANATOMY OF A SOLAR-ELECTRIC SYSTEM

OFF-GRID PV SYSTEM

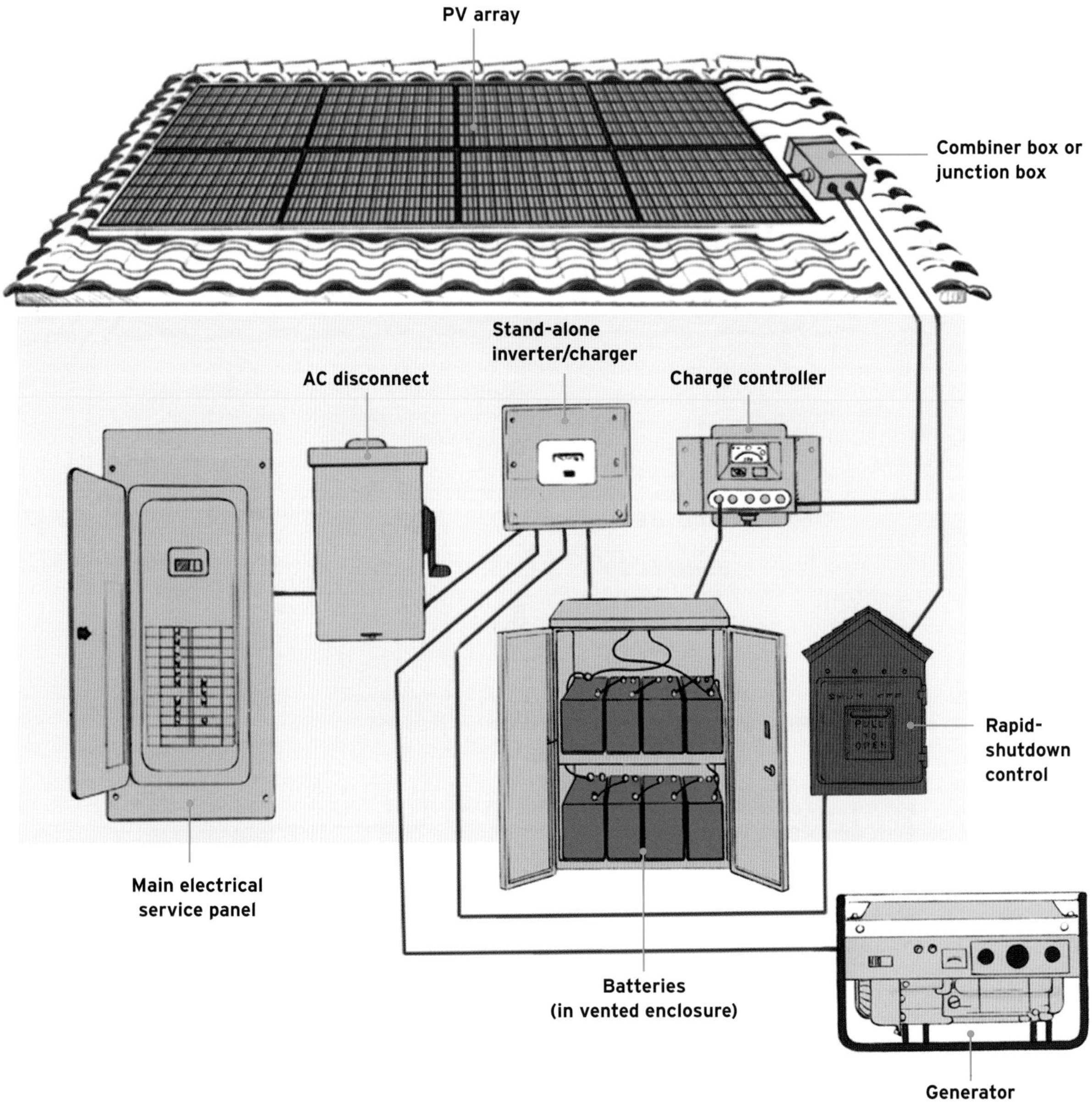

PV array
Combiner box or junction box
Stand-alone inverter/charger
AC disconnect
Charge controller
Rapid-shutdown control
Main electrical service panel
Batteries (in vented enclosure)
Generator

Off-grid PV systems are autonomous — or "stand-alone" — because they can store energy on-site. Most off-grid systems include a conventional generator to provide backup power when battery storage levels get too low and there is insufficient solar energy to recharge the batteries and/or accommodate the household usage.

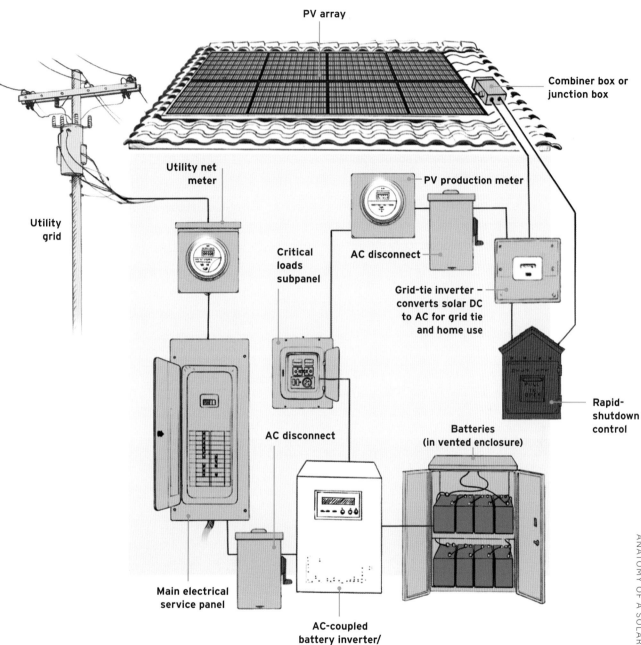

PV array

Combiner box or
junction box

Utility net
meter

PV production meter

Utility
grid

Critical
loads
subpanel

AC disconnect

Grid-tie inverter –
converts solar DC
to AC for grid tie
and home use

Rapid-
shutdown
control

AC disconnect

Batteries
(in vented enclosure)

Main electrical
service panel

AC-coupled
battery inverter/
charge controller

Grid-tied systems with battery backup can feed to, and draw from, the utility grid when the grid is operational. They can also store energy on-site for use when the grid is down or, optionally, during peak-rate periods when grid energy is the most expensive. The batteries can be recharged by the solar array as well as the grid. An AC-coupled system is shown here.

Grid-tied systems with battery backup are relatively complex, technically sophisticated, and pricey, costing significantly more than a standard grid-tied system. There are two main types of battery backup systems: DC-coupled and AC-coupled. DC-coupled systems are the historical standard, while AC-coupled systems are becoming more common and are the only type allowed by some utility companies, because they make it easier to track solar production.

A couple of important notes about grid-tied systems with battery backup:

1. Given the complexity of these systems, it's best to hire a professional for the system design and installation. (Only the installation of the array hardware is the same as that for the grid-tied and off-grid systems shown in this book.)

2. You can add battery backup to an existing grid-tied system, depending on the system type and design. If you install a standard grid-tied system now and later decide you'd like battery backup, you can easily add the components at a later time, provided you use an AC-coupled configuration. A DC-coupled system would likely require removing and rewiring the modules to accommodate the backup system's lower-voltage charge controller.

TIP
BACKUP BASICS

Grid-tied systems have the advantage and convenience of using the utility grid as a backup power source, so these PV systems do not have to provide all of the power to cover your household electricity usage. Off-grid systems are very different. Since there is no utility service to provide backup power, you must accurately size your PV array and battery bank for your household electrical usage.

YOU SAY "PANELS"; WE SAY "MODULES"

Here's a quick guide to basic solar-electric system terminology. Since you'll be working like a pro, you might as well sound like one.

PV: Photovoltaic or photovoltaics; the science of, or devices for, creating DC electricity with solar energy (i.e., photons, packets of light energy from the sun).

Module: A PV panel or collector. Pros use the word *module* when referring to PV systems to distinguish them from panels or collectors used for solar thermal systems for hot water or space heating.

Hardware: The main components of a PV system, including modules and their mounting structure, inverters, disconnects, meters, conduit, wiring, electrical boxes, and batteries, as applicable.

Mechanical: The phase of system installation that includes the module mounting structure, modules, and module wiring.

Mounting structure: The assembly of hardware that supports the PV modules and anchors them to the roof or ground.

Electrical: The installation phase for everything beyond the modules, including wiring runs, grounding, inverters, meters, disconnects, and batteries.

Can I Install My Own PV System?

A DIYer's Checklist

It's time for the litmus test that tells you whether to proceed boldly as an amateur solar installer or to hand over the reins to a professional. For most of you, the decision will come down to the rules of the local building authority (most likely your city, county, township, or state) or your utility provider, either of which may require that solar installations be done by a licensed professional. This is also the best time to confirm that your project won't be nixed by your zoning department, historical district standards, or your homeowner's association.

☐ Amateur installation is permitted by the local building authority and your utility provider.

☐ Requirements for amateur installation are reasonable and acceptable. Some authorities require nonprofessionals to pass tests demonstrating basic knowledge of electrical and other household systems, but such tests may not be extensive.

☐ You're okay with several hours of physical rooftop work (those with ground-mount systems get a pass here) AND you're wise enough to wear legitimate fall-arresting equipment (not a rope tied around your waist). You may feel as confident as Mary Poppins dancing on rooftops, but she can fly; you should be tethered.

☐ You don't live in a historical district or, if you do, the zoning authority permits PV systems (with acceptable restrictions).

☐ Your homeowner's association, if you have one, permits PV systems (with acceptable restrictions). Sometimes the homeowner's association may need a little nudging to give permission.

☐ You have a standard type of roofing (asphalt shingles, standing-seam metal, wood shingles, standard flat roof). If you have slate, concrete tile, clay tile, or other fragile/specialty roofing, consult a roofing professional and/or hire out the PV installation (see Don't Have Asphalt Shingles?, page 42). This is not necessarily a deal-breaker.

WARNING: PV systems are inherently dangerous and potentially deadly. As a DIY system installer and owner, you must understand, respect, and mitigate the risks involved with all installation and maintenance tasks. Pay special attention to the safety warnings given throughout this book as well as all requirements in the local building and electrical codes and equipment instruction manuals.

The installation phase of a typical residential grid-tied system takes pros about two to five days to complete, depending on the system size. It might take you and your helpers a day or two more. That's the good news. The bad news is that the entire process can easily take a few weeks and often stretches out to a few months. Why? Red tape, of course. While some communities have solar-friendly policies and procedures, many remain resistant to streamlining the approval process. Take heart, and rest assured that they, like Copernicus's detractors, must someday accept the importance of the sun.

Goals

Look at your electrical power usage and consider your power production goals, budget, and likely location for the PV array.

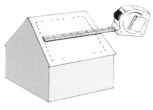

2 Site Assessment

Determine your system size (solar-electric production goal). Measure and map array installation area.

3 Design

Design the entire PV system according to your production goal. Choose modules and other main components. Complete a system plan for yourself and required documents for permitting.

4 Approval

Obtain permits from building and zoning departments, and approval from your homeowner's association or other authorities, as needed. Some utilities require an interconnection agreement as part of the approval process.

5 Shopping

Buy your PV system hardware (online or through local distributors or vendors) and have it shipped to your door. Get standard wiring, fasteners, and other basic supplies at local retailers.

6
Mechanical Installation

Install your module support structure (rooftop racking or ground-mount structure) and PV modules. Install microinverters, as applicable. Rough inspection by a city inspector typically happens here, usually *before* the module installation is complete.

7
Electrical Installation
(Preliminary)

Run conduit between your array and the ground-level components. Install a mounting bracket or box for a string inverter, as applicable.

8
Electrical Installation
(Final Connections)

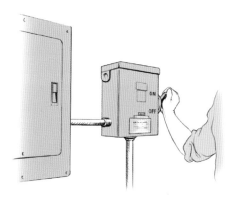

This step is for your electrician. He or she will pull wires; install the string inverter, AC disconnect, and production meter; and make all final AC electrical connections for the entire PV system — up to your home's main electrical service panel.

9
Final Inspection and Utility Hookup

Pass the final inspection by the city inspector, and bring in the utility worker to install the net meter and make the connection to the power grid.

10
PTO

Obtain your Permission to Operate (PTO) letter from the utility. Then turn on your PV system and start using solar-generated power!

Working with Solar Professionals

Even if you're planning to do as much of the installation yourself as possible, you'll need an electrician for everything on the AC side of the system, including the final hookup to the house electricity. You may also need to hire a solar professional, architect, or engineer to approve your design, especially if such approval is required for a permit. For those of you who have abandoned the DIY option, an experienced solar installer is the best way to go. Here is a basic rundown of the solar pros who can advise and assist you in going solar.

Meet the Solar Pros

Electrician. A licensed electrician *experienced with residential PV installation* can perform the final electrical installation as well as help with any other aspect of the electrical design. If you have experience with electrical equipment installation, you might negotiate a plan wherein you install boxes, run conduit, and do other grunt work, and your electrician pulls all the wiring and connects all the components. If you're a beginner, turn everything over to your electrician after the array is installed. In any case, all AC connections must be made by a licensed electrician. Contract with your electrician during the design phase so all parties know what to expect before installation begins.

Architect/engineer. Architects and engineers can help with structural design considerations and specifications, such as roof attachment, roof loads, ground-mount structure design and installation, and specialty mounting systems, like those for flat roofs. The local building authority might require an engineer's stamp on any installation plan. Review and approval of detailed plans might cost several hundred dollars, while design services and expert advice might be charged at an hourly rate.

Solar w/consultant. A DIYer might hire a local solar installer to consult on the design, provide technical help, or purchase system equipment at a competitive price. If you'd like to get this kind of à la carte help from an installer or company, just keep in mind that the installer's time is money, just like everyone else's. It's fine to call up one or more installers and ask about their services and discuss how they might help with your project, but resist the urge to pump them for free advice; their expertise is valuable.

Solar installer. Solar installers offer turnkey planning, design, and installation services. They also apply for rebates that lower your cost and advise you about tax breaks and other incentives that homeowners must apply for themselves (again, always check with your tax adviser). Good installers offer competitive warranties for their work and system components, and many are happy to provide technical assistance down the road (within reason, of course).

If You Don't DIY: Tips for Choosing a Solar Professional

Choosing a solar professional to install your system is, in many ways, a lot like finding any other home contractor, and the basics of due diligence certainly apply: get more than one bid, ask for references, and look at the company's history and reputation. But given the relative youth and rapid growth of the industry – and the fact that much of your investment will be tied up in the solar equipment – it's especially important to check for expertise, experience, and solid products and warranties.

◇ **Look for expertise with PV systems.** Even if a company is relatively new, it should have experienced installers in charge. At least one member of the installation crew should have PV Professional Installation certification from the North American Board of Certified Energy Practitioners (NABCEP). You can ask for proof of NABCEP certification, as well as the license number of the electrician who will perform the final AC hookups. Installation companies commonly have to provide these when bidding for certain jobs.

◇ **Check the member directories** of local industry associations if you need a place to start your search. The Solar Energy Industry Association (SEIA) and the American Solar Energy Society (ASES), two of the biggest national groups, sponsor many state and regional chapters. For example, the Colorado chapter of SEIA is called COSEIA; California's is CALSEIA.

◇ **Expect turnkey service.** Design, installation, permits, paperwork, rebates, utility hookup – the works. You shouldn't be asked to do anything a pro normally does.

◇ **Data monitoring.** Most inverters include monitoring capability (hardware and software) that allows you to check your PV system's performance from a computer or other device. Your solar professional should know how to hook up this function and teach you how to use it.

Answering the Big Three

When you call a professional solar installer to inquire about adding a PV system to your house, before long you'll be talking about three things: (1) area, (2) electrical loads, and (3) budget. As a DIY installer, you should ask yourself the same questions to confirm that both you and your house are ready for the project ahead.

1. Area: Where Are the PV Modules Going?

On a typical home in the Northern Hemisphere, the ideal place for a PV array is a south-facing rooftop. Here, the modules are out of the way (and not taking up space in your yard), they face the sun during the peak hours of daylight, and they're above a lot of obstructions that might cast shadows onto the array. It's okay if your roof doesn't face due south; PV systems can work on east- and west-facing roofs, too, but any rotation toward the south is helpful.

To determine whether your roof is big enough, start with a ballpark estimate: PV modules can produce about 10 to 15 watts per square foot of area. A 5,000-watt, or 5-kilowatt (kW), system needs about 500 square feet of roof area. (We'll get to sizing in a minute.) You can probably take quick measurements of your roof area from a ladder (if not, get on the roof), but keep a couple of things in mind:

◇ You can't install modules within 5 to 10 inches of any roof edge or the roof peak. In some jurisdictions this no-installation zone may extend as far as 3 feet from the edge or peak; ask the local building authority for guidelines.

◇ Modules cannot cover most roof penetrations, such as plumbing vents, chimneys, and skylights. However, sometimes obstructions can be cut shorter or trimmed to allow modules to be installed over them. Consult the local building authority for recommendations.

Ground-mounts can be located and positioned for optimal sun exposure.

If your roof turns out to be a poor candidate due to space constraints or lack of sun exposure (see Shade Matters, page 32), you can set your sights on a ground-mount array, which is not a bad thing. Ground-mounting adds some cost and time to the installation but offers a few key advantages: you can position the array for optimal sun exposure, you don't have to drill holes in your roof, and you never have to remove the array to replace your roofing. What you do need is a sizable space in your yard or on your property that's completely or nearly free of shade throughout the year.

TIP

BUILD FOR SOLAR

If you happen to be building a new house or large addition, try to keep south-facing roofs free of penetrations, leaving a wide-open space for a sizable PV array.

2. Loads: How Much Electricity Do You Need?

Answer this question by determining how much electricity you use. This is as easy as adding up a year's worth of electricity bills. However, looking at two or more years is better, especially if you have air-conditioning or any electric heating, as these are more weather-dependent than lighting and regular appliance use. Just be sure to use only full years for accuracy, and to look only at *electrical* usage, not natural gas or heating oil.

Gather your utility bills for 12 consecutive months and add up the total kilowatt-hour (kWh) usage of each month to determine your annual usage. Many utility bills now have a bar chart or graph showing your monthly electrical usage for the entire year on a single bill. If you look at more than a year's worth of bills, add up the total from all complete years and divide by the number of years to find the average annual usage. You will use this number for comparison when you calculate your system's estimated solar production with PVWatts, in chapter 2.

ANSWERING THE BIG THREE

17

Note: Many utility bills now have a bar chart or graph showing your monthly electrical usage for the entire year on a single bill. In case you're curious, the national average is about 900 kWh per month (most homes range from 400 to 1,000 kWh/month), or about 11,000 kWh per year. That's a pretty high number, but if you don't heat with electricity or cool your house with conventional air-conditioning, your usage is probably well below average. Of course, any significant energy efficiencies you can employ now – whether it's replacing old incandescent lights with LEDs, unplugging an extra fridge in the garage, or swapping your air conditioner for a whole-house fan or evaporative cooler – will allow you to get by with a smaller and less expensive PV system.

3. Budget: How Much Can You Spend?

For general reference, a PV system with a DIY installation can cost about $2 per watt of capacity. Professional installation can typically cost anywhere from $3 to $5 per watt. Residential systems typically range in size (or output) from about 3 to 8 kilowatts. That means a total initial cost of about $6,000 to $16,000 for a DIY system and about $15,000 to $30,000 for a professionally installed system. Smaller systems tend to cost more per watt than larger systems, due to economies of scale.

Regardless of who installs your system, the federal tax break available through 2021 can reduce your total cost by up to 30%. This is a tax credit subtracted from the tax amount you owe; it is not a check from the federal government. Additional rebates and incentives offered by utilities or state or local programs can reduce the cost further. See Solar Financials (page 94) for more information on financial incentives.

As you've probably heard, rebates and other incentives seem to be a dying breed, but don't feel too bad. Today the cost of a PV system is about one-third of what it was in the year 2000. And if you installed a PV system back in the 1970s, you might have paid anywhere from $12 to $50 per watt.

After answering the Big Three questions (or at least discussing them), a pro's next step is to visit your house to take measurements and perform a more thorough assessment of your site. You will do the same – and that's the focus of chapter 2.

TIP

ROOFING REPLACEMENT

PV modules can easily last over 30 years, which means they'll likely outlive your roofing material. And a new roof means temporarily removing the modules. Not a huge deal, especially if you installed them in the first place. But it's not something you'll want to do after just a year or two. So if your roof is ready for replacement, you should think seriously about tackling that job before adding a PV system.

VOLTS, AMPS, WATTS, AND WATT-HOURS

Electricity is the result of electrons flowing through a conductor, such as wires in a standard electrical circuit. All electrical circuits have a measure of voltage (or volts) and amperage (or amps). Simply put, **voltage** is the measure of pressure in a circuit, the force that pushes the electrons through. **Amperage**, also called the **current**, is the measure of the flow of electrons in a circuit. Together, the voltage and amperage tell us how much electricity is moving through the circuit, as well as how much electricity a circuit can safely handle.

A **watt** is an instantaneous unit of power that allows us to calculate the usage of an electrical device. It is derived by multiplying the voltage and amperage in an electrical circuit (volts × amps = watts).

A 10-amp appliance that plugs into a standard 120-volt household circuit has a wattage rating of 1,200 (10 × 120 = 1,200). This rating tells you how much wattage the appliance uses under normal operating conditions. Some appliances, like refrigerators and air conditioners, use more than their rated wattage when their motors are starting up.

However, we are interested in *energy* usage, which is power over time. To find the energy usage, simply take the power rating of the device and multiply it by the number of hours the device is used:

A 10-watt lightbulb running for 1 hour uses 10 **watt-hours** of electricity.

A 100-watt fan running for 10 hours uses 1,000 watt-hours, or 1 kilowatt-hour (kWh), of electricity. Thus, kilowatt-hours = watt-hours divided by 1,000.

ANSWERING THE BIG THREE

19

2 Assessing Your Site

GOAL

Confirm the size and location of your PV array

NOW THAT YOU'VE TAKEN THE FIRST STEPS of estimating your energy needs and thinking about where your PV system might go, it's time to take a closer look at both to determine the best location for the array and, ultimately, how big it needs to be. You'll determine the size and output with PVWatts, the free online calculator provided by the National Renewable Energy Laboratory (NREL). PVWatts uses your home's location and several other factors to estimate the amount of electricity you can expect to produce with a given system size. What it can't do is visit your site, so it's your job to measure your installation area, your roof slope, and your azimuth (not as bad as it sounds) and to consider additional factors, like shade, to come up with an accurate, realistic system size.

Measuring Area, Slope, and Azimuth

Area, slope, and azimuth are the measurements that describe the placement and position of your future array. They are the primary design factors that are specific to your site, and you will need them in order to use the PVWatts calculator.

Area: The usable square footage of roof space or ground area for installing the PV modules.

Slope: The angle, or pitch, of your roof. The slope determines the tilt of the modules.

Azimuth: The direction your roof or ground-mount array faces (south, southeast, west, etc.), measured in degrees.

With a rooftop system, all three of these factors are limited by what you already have or are planning to build. With a ground-mount system, you get to choose the slope (the module tilt angle) and azimuth, provided there is sufficient area in a suitable location with minimal shade.

How to Measure Area

Accurate measurements are easiest with a helper and a 25-foot or longer tape measure. If you like, you can record your measurements on a simple map. The following steps apply to rooftop arrays. For ground-mount arrays, simply measure the available open space in the most likely installation area of your yard or property.

1. Before you climb onto the roof, draw a simple aerial-view map of any roof areas suitable for PV modules; this is usually a south-, southeast-, or southwest-facing roof area. Note on your map the relative locations of penetrations and obstructions, such as roof vents, plumbing pipes, skylights, and chimneys.

2. Once you're on the roof, measure the length and width of each roof plane on your map. A suitable space typically measures at least 15 feet from the eave to the roof peak (also called the ridge) and 20 feet from side to side, but it's possible to install a small grouping of modules in a smaller area.

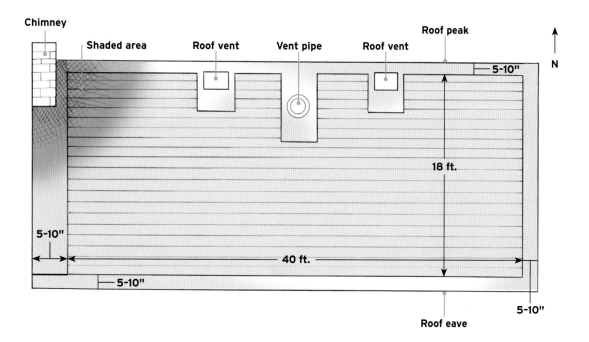

3. Subtract 5 to 10 inches from the installation area for each roof edge and the ridge to allow for the required minimum margin around the array. Remember, your local building authority may have different requirements, such as a 2- or 3-foot-wide traffic path for firefighters; the 5- to 10-inch margin at the ridge and edges prevents wind uplift on the modules.

4. Measure and record the locations of all obstacles and penetrations. Include a margin of a few inches around all sides of each obstacle. **Note:** If an obstacle will cast a shadow over any of the installation area during peak sun hours, you may need to omit the shaded area from your total (see Snow, Wind, and Shade, page 31).

5. Note the dimensions of specific installation areas, particularly if the space is chopped up due to penetrations and obstructions. You will use these dimensions to determine how many modules you can fit in specific areas.

How to Measure Slope

There are several methods for measuring roof slope. If you have a smartphone, see if it includes a level app (measures for levelness). This may be part of the compass app, which you'll use to measure azimuth. Open the level app; set the phone, on edge, on the roof; and read the degree of angle given.

Another method is to use an inexpensive tool called an inclinometer. This works like an analog version of the phone app: set it on the roof and read the degree angle indicated by the arrow.

Then there's the old-school method, which requires a little background. In builder's parlance, the slope (or pitch) of a roof plane is expressed in a rise-run ratio. A 5-in-12 (or 5:12) roof pitch rises 5 inches for every 12 inches of horizontal run. The chart below shows that a 5:12 roof is angled at 22.6 degrees.

You can measure the rise and run with a framing square (carpenter's square) and a small level, such as a torpedo level or 2-foot level (see the image on the facing page).

1. Position the square against the roof edge so that the edge intersects the 12-inch mark on the long, horizontal leg of the square.

2. Using the level as a guide, adjust the square so that the horizontal leg is perfectly level.

3. Note where the roof edge intersects the vertical leg of the square; this is the rise value. If the vertical leg shows 5 inches, you have a 5-in-12 roof slope.

4. Convert the rise-run slope to degrees, using the chart. You can also find conversion charts online, for example, at PVWatts (click on the "i" in the "Tilt (deg)" row on the "System Info" page). Or you can use a scientific calculator: the degree angle is the inverse tangent of rise/run: divide the rise by the run, then press the inverse tangent button on your calculator.

ROOF SLOPE TO DEGREES

SLOPE	3:12	4:12	5:12	6:12	7:12	8:12	9:12	10:12	11:12	12:12
DEGREE ANGLE	14.04	18.43	22.62	26.57	30.26	33.69	36.87	39.81	42.51	45.00

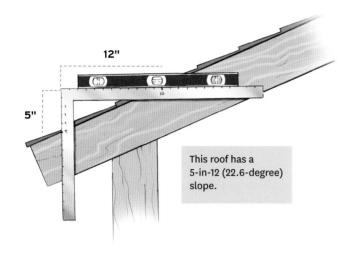

12"

5"

This roof has a
5-in-12 (22.6-degree)
slope.

How to Measure Azimuth

Azimuth is the compass direction the PV array faces. For most rooftop arrays, this is also the direction the roof faces. An array facing due north has an azimuth of 0 degrees; due east is 90 degrees, due south is 180 degrees, and due west is 270 degrees. You can measure azimuth with a smartphone or an old-fashioned compass. Smartphones are easier because they calibrate themselves to your location so you don't have to account for magnetic declination (the variation between true north and magnetic north). Compass apps also work on tablets, and you can download free versions for any compatible device.

To use a compass app, simply open the app and follow the instructions for calibration. When the app is ready, stand with your back to the roof where the array will go, and hold the phone flat in your hand and directly in front of you. The app gives you the precise angle the roof plane faces; this is the azimuth for your roof.

You can use an old-fashioned compass the same way. Just remember to account for magnetic declination for your location. If you don't know that information off the top of your head, you can look it up with the online calculator provided by the National Oceanic and Atmospheric Administration (NOAA). Also, make sure there are no metal or other

TIP

DON'T HAVE A FRAMING SQUARE?

If you don't have a square, use any level that's at least 12 inches long (or attach a small level to a foot-long board) and a tape measure. Mark the level 12 inches from one end. Position that same end of the level against the roof and hold it level. Measure straight down from the 12-inch mark to find the rise.

FLAT ROOFS WORK, TOO

PV arrays on sloped roofs typically mount to a "flush-mount" (parallel to the roof slope) racking system that is anchored into the roof framing. Arrays on houses with flat roofs use special supports that are held in place by weights (called ballast); the arrays are usually not fastened to the roof. Mounting structures are available with a variety of tilt angles, typically ranging from about 10 to 20 degrees and almost always less than 20 degrees. Modules at steeper angles are subject to higher wind loads and often must be mechanically anchored to the roof. Otherwise, the higher the tilt and/or wind loads, the more ballast is required. See Mounting Structures for Flat Roofs (page 44) for more information.

25

magnetic objects nearby, as this can throw off the compass needle. Don't forget your metal watch or belt buckle!

For those of you planning a ground-mount array, it's generally best if the array faces due south, an azimuth of 180 degrees.

Sizing Your System with PVWatts

Just for fun, take a minute to go online to the PVWatts calculator (pvwatts.nrel.gov). Enter your city and state, then click "Go." On the next page, click on the map location closest to your house, or stick with the default location selected for your city, then click on the big orange arrow labeled "Go to system info." On the next page, click on the big orange arrow labeled "Go to PVWatts results." Boom. The number in large type at the top is how much AC electricity you can produce per year at your location with a 4 kW (DC) system facing due south. The system size and other default inputs are already in the calculator.

Of course, your system may not face due south and it may not be 4 kW, but this 60-second exercise shows you how easy it is to start using PVWatts. No log-in or registration required, no cost, and no ads for treating toenail fungus. Now, that's the government doing something right.

PVWatts is incredibly simple on the surface, but there's much more you can do with it to improve accuracy. It's designed for trial and error, so you can easily click back and forth between the pages, plug in different data, and immediately review the results. The "System Info" page is where you'll enter your specific data.

System Info Parameters

The PVWatts "System Info" page includes six basic design parameters for sizing your PV system. We'll briefly discuss each one here,

and you can refer to PVWatts for more detail, as needed; just click on the "i" button to the right of each parameter. There's also a button labeled "Advanced Parameters." You can ignore this for now and rely on the system default inputs; chances are, you'll never need to tweak these values.

DC SYSTEM SIZE

System size is the rated output of DC power for the entire solar array (see STC Ratings, page 31). A system with an array of 20 PV modules rated at 250 watts each is a 5 kW system (20 × 250 watts = 5,000 watts). The default system size on PVWatts is 4 kW, but you can change this to any size you like. Don't worry about entering a size at first; when all of your other parameters are in place, you can play around with different sizes and quickly check the results until you find a system size that yields your target power output.

Note that your target output is in AC and represents the amount of usable electricity produced annually by the system. The DC system size is not the same thing as the annual AC production. It's easy to get confused here because, depending on where you live, the DC system size and the annual production may have similar numbers attached to them. For example, a 5 kW (5,000-watt) system may produce around 5,000 kilowatt-hours (kWh) of electricity annually. But for all intents and purposes you can consider this a coincidence. A 5 kW system is capable of producing 5,000 watts of DC electricity at a given moment under ideal conditions and peak sunlight. How much usable energy the same size system can provide your home each year is based on many factors, and that's what PVWatts calculates for you. Using your location and your roof's (or ground-mount's) azimuth and tilt, it estimates how much sunlight your system will receive year-round and therefore how much electricity it is likely to produce.

MODULE TYPE

This parameter lets you select one of three types of PV modules: standard or premium conventional crystalline-silicon (c-Si) modules or thin-film modules. Conventional c-Si modules are typical for home power systems, so you'll likely either use standard modules (with an efficiency of about 15%) or you'll pay more for premium modules (with efficiencies up to about 20%). If you know the exact modules you'll be using, you can refer to the manufacturer's data sheet to help choose standard versus premium. Otherwise, use the default type: standard. If you are using modules made from a-Si, CIGS, or CdTe solar cell material, then you will select "Thin-film" as your module type.

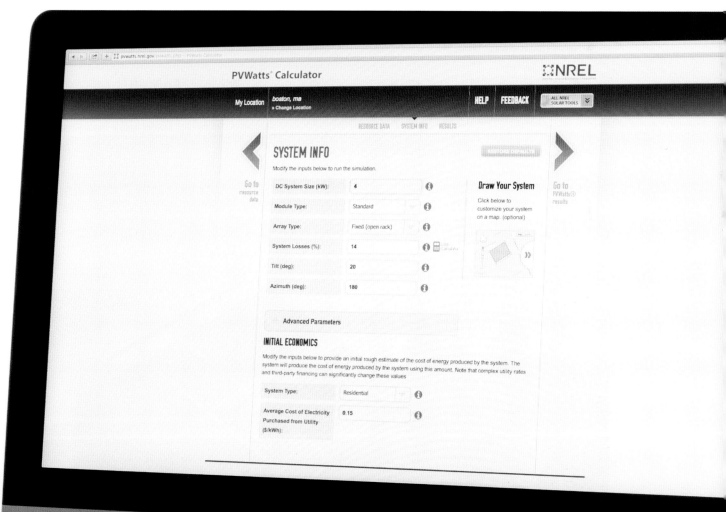

ARRAY TYPE

The array type is the module-mounting type: rooftop or ground-mount. If you're installing a standard flush-mount rooftop system, select "Fixed (roof mount)" in the drop-down menu. If you're doing a ground-mount system or a flat-roof system (where the modules are tilted up on a flat roof surface), select "Fixed (open rack)," unless you're keen on using a tracking system (see What's Tracking?, below). The fixed (open rack) option gives you a slight performance advantage due to the lower average temperature of a ground-mount (or flat-roof) array versus a flush-mount rooftop system. Most PV modules are slightly less productive in temperatures over 77°F (25°C), a temperature that's easy for the solar cell material to hit on a rooftop. If you are using tracking, choose either "Single-axis" or "Dual-axis" for your array type.

WHAT'S TRACKING?

Tracking is usually an option for ground-mount systems only. It works by way of a motor or other device that automatically moves the PV array throughout the day to maximize solar irradiance (how much sun is captured by the modules). *Single-axis* tracking rotates the array from east to west, following the sun across the sky over the course of the day. *Dual-axis* tracking rotates *and* tilts the array so that it always faces the sun directly.

Residential systems with tracking typically involve single-pole mounts holding about 8 to 12 full-size modules. Larger systems require multiple mounts, each with its own tracking device. Given the added cost of tracking, most home power systems use a fixed ground-mount instead. Tracking also introduces moving parts and their maintenance needs. If you're interested in tracking, it's a good idea to consult a solar pro for design and installation advice.

Some fixed ground-mount array supports are adjustable and can be tilted manually to face the sun in different seasons. This offers some of the benefits of tracking without the added cost and maintenance issues. Alternatively, if you have the space, simply adding a few more modules to a fixed ground-mount array may offer the same output boost you would get with a tracking system but would do so more cheaply and, again, without moving parts and added maintenance.

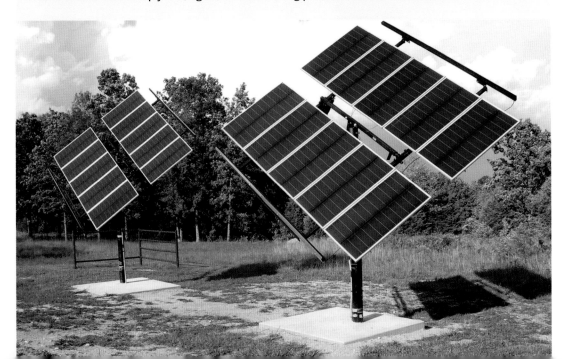

SYSTEM LOSSES

In short, system losses are the total of all the little reductions in efficiency due to equipment inefficiencies and real-world operating conditions. The losses account for the difference between the maximum rated DC power of the array and the actual AC power that is fed to your house or the utility grid.

The default system loss value in PVWatts is 14%, giving you an overall system efficiency of 86% (100% – 14% = 86%). This means that if your array is rated for 5 kW of DC power, it would produce 4.3 kW of usable AC power (5,000 watts × 0.86 = 4,300 watts). However, we recommend a somewhat higher loss value of 18%; see the PV System Losses chart (below) for a breakdown. This is based on decades of

experience with PV systems. Alternatively, you can learn about each loss category in PVWatts and enter your own values into the loss calculator to come up with a system losses percentage.

TILT

Tilt is the angle of your PV modules, measured in degrees: 0 degrees is horizontal; 90 degrees is vertical. For standard flush-mount installation on a sloped roof, enter your roof slope in degrees. For a ground-mount system, remember that you get to choose the tilt. A tilt equal to the latitude of your location is the standard compromise for year-round production. (A lower tilt yields higher production in summer; a steeper tilt is more productive in winter.) As an example, Denver, Colorado, is at 40

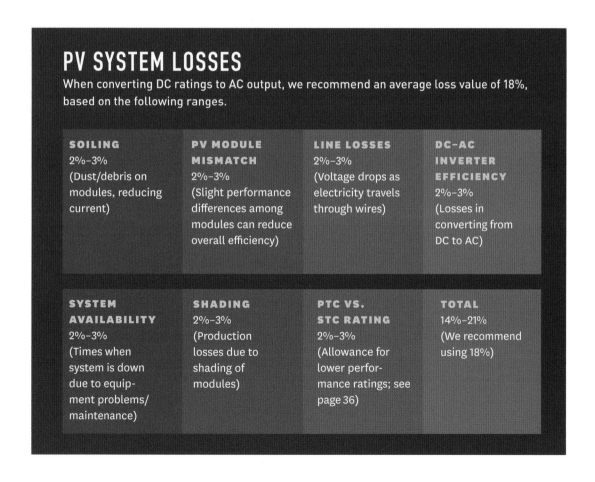

PV SYSTEM LOSSES
When converting DC ratings to AC output, we recommend an average loss value of 18%, based on the following ranges.

SOILING	PV MODULE MISMATCH	LINE LOSSES	DC–AC INVERTER EFFICIENCY
2%–3%	2%–3%	2%–3%	2%–3%
(Dust/debris on modules, reducing current)	(Slight performance differences among modules can reduce overall efficiency)	(Voltage drops as electricity travels through wires)	(Losses in converting from DC to AC)

SYSTEM AVAILABILITY	SHADING	PTC VS. STC RATING	TOTAL
2%–3%	2%–3%	2%–3%	14%–21%
(Times when system is down due to equipment problems/maintenance)	(Production losses due to shading of modules)	(Allowance for lower performance ratings; see page 36)	(We recommend using 18%)

degrees latitude, so the standard tilt angle for ground-mount systems is 40 degrees; however, a tilt anywhere from 30 to 45 degrees will work fine in Colorado. You can look up the latitude of your location with an online latitude/longitude finder.

You can also plug in different tilt angles in PVWatts and compare the results. Note that the "Results" page gives you the array output for each month of the year, allowing you to see how a change in tilt affects output during different seasons. Also note that while the summer and winter values may change significantly as you vary the tilt, the total annual energy output may change by only a few percentage points. In other words, it usually balances out over the year.

AZIMUTH

Simply enter the degree value you calculated earlier. If you're planning a ground-mount system and have some flexibility in positioning the array, you can play around with different azimuth values to confirm the ideal position for your location.

Working with the Results

Once you've entered your basic parameters in the "System Info" page, click on the orange arrow labeled "Go to PVWatts results." The number you get is the average annual kilowatt-hour production (again, in AC power) you can expect with a 4 kW system (the rated DC system size), unless you entered a different system size. This may be too low to meet your electricity needs (since the national average is about 11,000 kWh per year). If so, click the left arrow to go back to the "System Info" page, and enter a higher number for the "DC System Size." Click the right arrow again for your results. Repeat as needed until you hit the magic number.

This is assuming your goal is to cover a typical year's total electricity usage with PV power. You can certainly choose to cover more or less than that, but there are some important considerations with either route. If you go with a smaller system that produces less than your overall usage, the difference (for grid-tied systems) will be made up by utility power. This alone is not a problem; in fact, *all* grid-tied systems pull utility power on and off throughout the year and even throughout most days of the year. But keep in mind that the more utility power you use, the more you'll be affected by rate increases over the years.

If you choose to install a bigger system that produces more power than you're likely to use, be aware that utilities commonly put a cap on the amount of excess power residential PV systems can produce. For example, you may be limited to 120% of your annual usage. So if your annual household consumption is 10,000 kWh, you can't install a system that's capable of producing more than 12,000 kWh per year. (Utilities are forced to play nice with renewable energy producers, but they don't have to be *that* nice.)

Here's another consideration with large systems: Don't assume that you'll get full retail value for excess power that you sell back to the utility. In fact, it's safer to assume that you'll get much less. For example, you might pay the utility $0.10/kWh for grid power, but the utility might pay you only $0.05/kWh for your solar-generated electricity. And, because the pricing program is controlled by the utility, there's no guarantee that you'll get the same price (or any payback at all) indefinitely. Buyback rates, rules, and regulations, often generically referred to as **net metering**, are the subject of ongoing discussion among consumers, utilities, and regulatory commissions throughout the United States.

Snow, Wind, and Shade

Designing anything for the outdoors comes with special challenges. In the case of PV systems, manufacturers have eliminated most of the usual concerns by making their modules and other exposed components tough and highly weather-resistant. So you don't have to worry about things like rain, hail, and of course sun exposure. But snow, wind, and shade are site-specific factors that must be considered.

Snow and Snowmelt

PV modules don't produce electricity when they're covered with snow. However, this is a minor concern in most snowy climates. For one thing, when it's snowing it's also very cloudy, so the modules wouldn't produce much power even without the snow blanket. When the sun comes out, snow cover on solar arrays tends to melt relatively quickly because the modules heat up with solar radiation. If you want to speed up the process by clearing off some of the snow, use a wood-handled broom or mop that won't scratch the module glass. The wood handle is for electrical safety. Do not use a shovel, which might damage the modules. Otherwise, just wait for the sun to take care of it.

STC RATINGS

The rating of a PV module refers to its performance under standard test conditions (STC). This is essentially a car-mileage test for solar panels (as you might guess, actual mileage may vary). Modules carry a nameplate listing several STC criteria, such as voltage, current, and wattage. You will refer to some of these nameplate values when choosing your modules and other equipment during the design phase. The basic system is based on STC ratings, but you'll also perform some calculations using "extreme" (non-STC) values.

For those who are interested in the scientific data behind STC, the testing conditions for measuring PV module performance include the following:

- Solar irradiance (1,000 watts of solar energy per square meter; 1,000 W/m²)

- Temperature (25°C/77°F)

- Solar spectrum (AM1.5 Global)

PHOTOVOLTAIC MODULE

Solar Module Type : JKM315P-72

Maximum Power	(Pmax)	315W
Power Tolerance		0~+3%
Maximum Power Voltage	(Vmp)	37.2V
Maximum Power Current	(Imp)	8.48A
Open Circuit Voltage	(Voc)	46.2V
Short Circuit Current	(Isc)	9.01A
Nominal Operating Cell Temp	(NOCT)	45±2℃
Maximum System Voltage		1000VDC
Maximum Series overcurrent protective device rating		15A
Operating Temperature		-40℃~+85℃
Application Class		A
Module Fire Performance		Type 1
Weight		26.5(kg)
Dimension		1956×992×40(mm)

STC: 1000W/m², AM1.5, 25℃

System Fire Class Rating: See Installation Instructions for Installation Requirements to Achieve a Specified System Fire Class Rating with this Product
The fire rating is Class C in Canada

⚡ WARNING

ONLY qualified personnel should install or perform maintenance work on these modules
BE AWARE of dangerous high DC voltage when connecting modules
DO NOT damage or scratch the rear surface of the module

For field connections , use 12 AWG wire insulated for a minimum of 90°C, rated for wet conditions and resistant to ultra violet radiation (where exposed)

c(UL)US LISTED   C E

If you have a ground-mount array, you don't want ground snow or drifts covering or shading your modules. Simply install your array so that the lower edge is higher than the typical snow level. If you get an unusually heavy snow or drifting, you can shovel away the ground snow and wipe off the modules.

Rooftop systems require a couple of special considerations. The first is weight. Your roof and module mounting structure must be able to support the extra weight of snow loads (measured in pounds per square foot) in your area. You can learn about snow loads from the local building department, and they will surely be part of your design and permit approval.

The second consideration is avalanche. A PV array is essentially a big, smooth panel of glass. A heavy blanket of snow can lose its grip during melting, and everything slides down at once, dropping onto a deck, roof, walkway, or whatever else might be waiting innocently below. The standard way to minimize this risk is to keep the bottom edge of the array 2 feet or so above the bottom edge of the roof. Ask the local building department for specific recommendations. Also, you can install snow guards along the eave to break up the snow after it slides off the modules.

Planning for Wind

Wind is a standard design load that is established by the local building authority. Whether you're installing a rooftop system or constructing a support system for a ground-mount, the building department will make sure your plan includes sufficient structural strength to resist wind forces at your location. Follow these specifications to the letter; you can imagine what might happen when a sail-like array isn't properly secured. Wind is the reason rooftop modules are kept at least 5 to 10 inches from the roof peak and all edges, and why ground-mount

structures are anchored in concrete or other equally secure foundations. **Note:** The local building department may require a letter with an engineer's stamp verifying that your PV design (and roof) meets the local requirements for both wind and snow loads.

Shade Matters

Shade has a compounded effect on PV module output. If a single leaf falls onto a module, shading even a single cell, the electrical output of that cell is reduced proportionately to the amount of shading. And because the cells are wired together in series, all of the other cells in that module have their outputs reduced equal to the reduction of the shaded cell, lowering the output of the entire module.

If the system has a traditional string-inverter setup, the problem gets worse. String inverters treat module strings as single units, not as individual modules. Therefore, shading of a single module results in a downgrade for the entire string, which might be up to half of your array. Newer string inverters have electronics that reduce shading losses. Additionally, most modules are now wired with bypass diodes, which bypass the shaded cell (or module), allowing the rest of the string to behave normally.

Microinverters treat modules individually, since each module (or module pair) has its own micro-inverter. This limits the effect of shading to a single module. Even so, if a tree limb or chimney casts a shadow across several modules, the production losses add up.

The bottom line is, shade is not good, so it's important to install modules in areas with minimal shading.

So how do you know when an obstacle will shade your array/installation area? Apply the

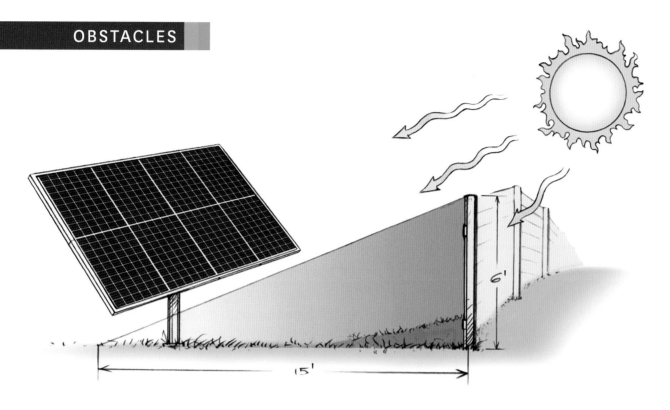

Obstacles can cast a shadow 2½ times their height when the sun is low in the sky. The higher the sun's angle, the shorter the shadow.

2½ times rule: *The length of a shadow can be 2½ times the height of the obstacle* (when the sun is at its lowest position in the sky). For example, a 20-foot-tall tree can cast a shadow up to 50 feet long, depending on the sun's position in the sky. (The longest shadows in the Northern Hemisphere occur on the winter solstice, between December 20 and 23.) With a rooftop array, you need to be concerned only with the relative height of the tree, that is, the portion of the tree that extends above the roof level. With a ground-mount array, you need to consider the full height of the tree, minus the array's distance above the ground. Also, with ground-mount arrays with multiple rows, you must space the rows so that they don't shade one another.

The easiest way to assess shading potential at your site is with your eyes: check the installation area at different times throughout the day (and throughout the seasons, if possible) to see whether, when, and where shadows appear. Combining visual inspection and the 2½ times rule is sufficient for most situations. For a more detailed assessment, some solar pros use a Solar Pathfinder, a simple but expensive instrument that is set up on-site and provides data on year-round sun exposure and potential shading. If you choose this route, it's probably best to have an expert do it for you. The Solar Pathfinder unit costs about $300, and while some installers swear by them, others find them too finicky or difficult to use and not worth the trouble.

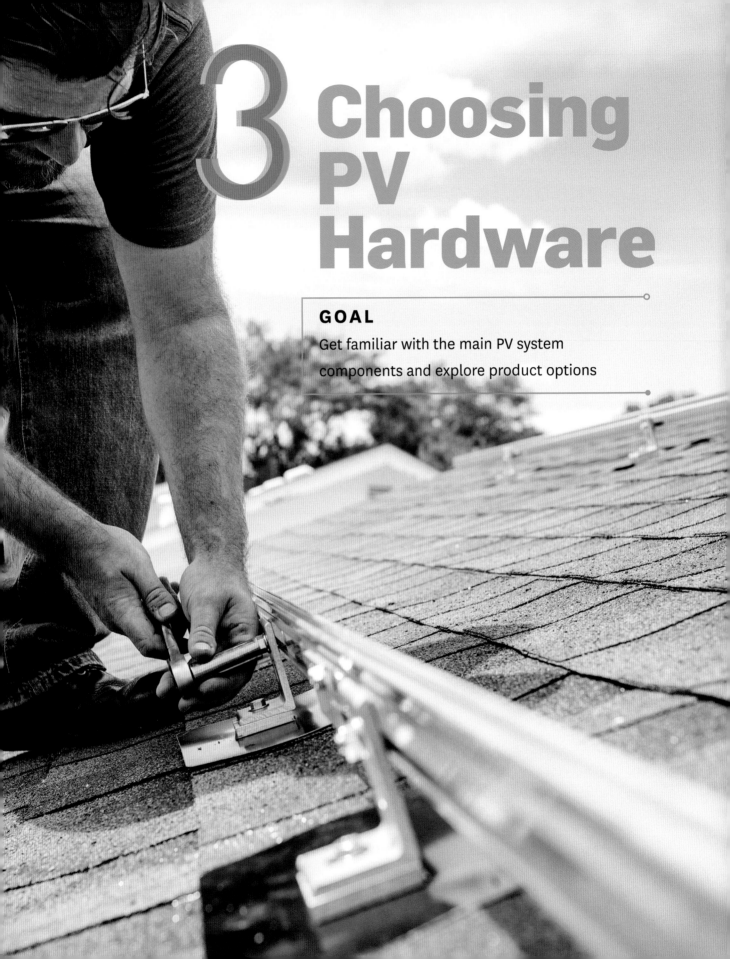

3 Choosing PV Hardware

GOAL

Get familiar with the main PV system components and explore product options

SHOPPING FOR PV EQUIPMENT is completely foreign territory for most people. There's a lot of technical data (not to mention geek chat) available for those who are interested, but for others this can be a little overwhelming. If you're in the latter camp, rest assured that much of the decision making is similar to that for any major household purchase, where you compare products for suitability, performance, value, and above all manufacturer quality and reliability. You'll also be glad to know that most of your purchasing decisions involve only three main components: the PV modules, the module mounting structure, and the DC–AC inverter(s). The rest of the parts needed for the installation, like basic electrical and construction materials, are relatively generic and can be purchased locally (see Local Materials, page 94).

At this stage of the project, you're just learning about the product categories and specifications and doing some online window shopping. Because the components' specs have to be right for your design, you won't finalize your product choices until the system design is complete (we'll get to that in chapter 4).

Modules

Modules are the heart of a PV system and represent the largest equipment cost, at least for grid-tied systems (batteries for off-grid systems can be equally pricey). The good news is that there are many, many options to choose from, and modules today are a steal compared to products from less than a decade ago. (Your neighbors who went solar about 10 years ago will be thrilled to hear that your panels cost about 75% less than theirs.) The following are the primary factors to consider when choosing modules.

Performance

The bottom line for PV module performance is power output. Output equals electricity production and is rated in watts. A module rated at 250 watts is capable of producing 250 watts of DC electricity at standard test conditions (see STC Ratings, page 31). There are a couple of additional ratings that shed more light on manufacturers' stated output ratings:

◇ **POWER TOLERANCE** is a rating that provides a range of how closely the module will perform to its rated value under STC. A power tolerance of +/- 5% means the module may perform 5% above or below its STC rating; for example, a 300-watt module may actually produce a low maximum value of 285 watts. Modules with a "positive-only" rating are rated to perform at or better than their STC ratings.

◇ **PTC RATINGS** are similar to STC but are based on a more realistic set of conditions, as determined by the test facility Photovoltaics for Utility Systems Applications (PVUSA), funded by the Department of Energy near Sacramento, California. PVUSA test conditions (PTC) use a higher cell temperature (113°F for PTC as opposed to 77°F for STC). Because crystalline-silicon cells are less productive at higher temperatures, PTC ratings are lower than STC ratings – typically 8% to 10% lower. You may find that a module's PTC rating is

given as a ratio: "PTC/STC." In this case the higher the number, the better. For example, a 250-watt (STC) module with a 0.9 PTC/STC ratio has a PTC rating of 225 watts (250 × 0.9 = 225). Another 250-watt module with a 0.85 PTC/STC ratio may produce only 212.5 watts (250 × 0.85 = 212.5). Keep in mind that PV system design is based on STC, not PTC, ratings. STC gives you the highest output possible, which is required for sizing the inverter, wiring, and other components. The GoSolar California website maintains a list of more than 1,600 models of PV modules and their PTC ratings.

All that being said, don't get too hung up on ratings. You will use the nameplate ratings of your modules to design your PV system, but the actual output on a daily, monthly, and annual basis will primarily depend on weather conditions such as sunlight, temperature, and precipitation. What a module does in the laboratory (under STC) is important for design and predicting system output, but once your system is in place, don't lose any sleep over it.

Manufacturer, Warranty, and Certification

PV modules typically are warranted separately for materials and power output. The standard for quality modules is 10 years on materials and 25 years on power output. If a manufacturer offers anything less, look for another module.

Warranties for power output typically state that the module will lose no more than 15%–20% of its rated output over the warranty period. For example, if a module's STC output rating is 300 watts, it should be capable of producing at least 240 watts after 25 years. But as noted in Cell Efficiency (right), crystalline-silicon modules typically lose only about 0.1% of their output per year, or a total of 2.5% over 25 years.

Perhaps more important than the warranty is the stability of the company behind the guarantee. If the company goes belly-up during your warranty period, that's probably the end of your warranty, although some manufacturers offer noncancellable warranties that are backed by insurance policies. On the bright side, well-made PV modules are highly reliable – many early adopters are still using modules that date back to the 1970s, proof that modules can last well beyond their 25-year warranties.

The best manufacturers control production of their products from start to finish to ensure quality standards. They also invest in research and development and have a long-standing reputation to uphold. At the other end of the spectrum are companies – many of them relatively new – that are more module assemblers than manufacturers. They may use parts from a variety of suppliers and have only limited control over quality.

Any module you consider must carry the UL 1703 certification from Underwriters Laboratories. Its nameplate should carry the UL stamp and/or the International Electrotechnical Commission (IEC) stamp. The local building department may require additional certifications. Certifications typically are noted on the module spec sheet; if not, look on the manufacturer's website.

SOLAR CELL EFFICIENCY

One term you hear a lot when shopping for PV modules is **efficiency**. In this case, efficiency refers to the percentage of solar irradiance that is turned into electricity by the solar cells. Standard crystalline-silicon PV modules offer somewhere between 15% and 20% efficiency. Generally speaking, the best modules offer the highest efficiency, but plenty of good, well-priced products are at the lower end of the efficiency range. (Note: Though the numbers are similar, module efficiency is not the same thing as the overall system losses value you entered into PVWatts in chapter 2.)

It's easy to get caught up in efficiency ratings, but it's not worth the trouble. For one thing, it's difficult to precisely calculate cell efficiency in practical use, so a percentage point here or there isn't something to base your budget on. And second, just because a module can't use 85% of the energy that strikes it doesn't mean that energy is wasted. Solar energy is totally free and inexhaustible. You *can't* waste it. That makes for a scenario completely different from, say, a 90% efficient gas water heater, which wastes 10% of the gas it burns due to heat losses. You're paying for that gas, and there's an environmental cost to using it. Not so with solar energy.

Solar cell efficiency degrades over time, depending on the photovoltaic material, which means modules produce less electricity as they age. Most module warranties are based on this. A typical warranty might guarantee an efficiency loss of no more than 15%–20% over 25 years. However, this is an extremely wide safety margin for the manufacturers. In reality, quality crystalline-silicon modules lose only about 0.1% efficiency per year.

Monocrystalline or Polycrystalline?

Conventional crystalline-silicon (c-Si; sometimes called x-Si) modules are made with one of two types of silicon: monocrystalline or polycrystalline. While the performance and costs of the two types of modules have largely equalized over the years, you still have to choose between the two. Here are the highlights:

MONOCRYSTALLINE CELLS . . .

◇ Typically offer slightly higher efficiency, sometimes resulting in slightly smaller modules with equivalent output.

◇ Typically have angled corners due to manufacturing processes.

◇ Have consistent dark coloring.

◇ Cost more historically, but prices are now comparable.

POLYCRYSTALLINE CELLS . . .

◇ Have square corners; this means that more of the module area is covered with cell material, which helps to counteract lower efficiency.

◇ Often have a blue tint and a mottled appearance.

◇ Usually have a "P" in their model number to indicate polycrystalline.

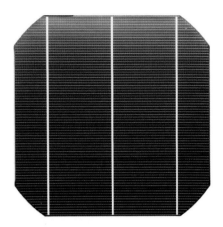

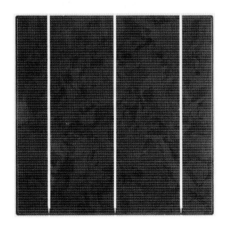

Size

Module size usually comes with a trade-off. A larger module may have higher output, but it's also heavier and more expensive, and of course it takes up more space than a smaller module. On the other hand, higher output might mean you can use fewer modules: if you need a 3 kW system, you can use 10 modules of 300 watts each as opposed to 12 modules of 250 watts each, provided your site can accommodate the larger units. Or you might have a small roof area that can accommodate two large modules but not three smaller ones. Manufacturers' spec sheets provide accurate dimensions for modules, but the standard module width is about 39 inches; 60-cell modules typically are about 65 inches long, and 72-cell modules are about 77 inches long.

Color

Standard modules have silvery aluminum frames and white backing material, or back sheets. Also available are units with black frames and white back sheets, or all-black units with black frames and black back sheets, which you might pay a small premium for. The black may blend in better on some roofs. Color is primarily an aesthetic consideration and has little effect on module performance, except that modules with black back sheets may run a few degrees warmer, slightly reducing the module's efficiency.

Price

Price speaks for itself, but as you've probably guessed, it's not a good idea to choose modules based on price alone. Once you've settled on a good fit for your system and budget, you can check on options for the best price. For example, sometimes you can save by buying a whole pallet of modules, which can reduce shipping cost and/or the price per module. This works well if the pallet quantity matches what you need or if you can go in with a friend or neighbor and share the modules from a large pallet.

Some vendors provide a $/watt (price per watt) value, which is handy for comparing modules. This can be calculated by dividing the module's price by its nameplate power output. For example, a 250-watt module priced at $250 costs $1/watt.

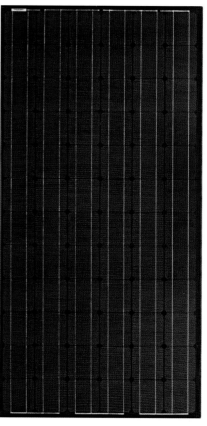

All-black PV module

Standard PV module with aluminum frame and white backsheet

Module Support Structures

Support systems for PV arrays are the metal structures that hold the modules, typically in a flat plane, and anchor the array to the roof or the ground. System options are based on roof slope (and roofing material) or ground-mount design. Flat roofs typically use systems that are secured by weights (called ballast), not screws or other anchors.

Mounting structure specs are required for most PV installation permits. This ensures that the system is suitable for the conditions in your area — snow loads, wind, coastal air, soil conditions (for ground-mounts), and so on. An experienced PV equipment supplier can recommend product ratings for your climate, or you can get the required specs from the local building department. You also must confirm that your roof can support the weight of the racking and the modules. This may involve simply answering some questions on your permit application, or it may require an examination and approval by an engineer. Keep in mind that conventional modules and racking typically add a weight load of 3 to 5 pounds per square foot.

Racking for Sloped Roofs

Module supports for roofs are commonly called **racking**. The standard racking style for sloped roofs is a **flush-mount** system with horizontal rails that support and secure the modules above and are mounted to the roof below with heavy-duty brackets, often called **footers**. The footers are anchored into the rafters or trusses of the roof frame with lag screws. On shingle roofs, each footer is coupled with a piece of metal flashing to help prevent leaks and damage to the roofing material. Some footers come with integrated flashing.

Because crystalline-silicon modules lose efficiency at higher temperatures, all flush-mount PV arrays are elevated several inches above the roof to provide cooling airflow underneath the modules (see Critter-Proof Your Array, opposite).

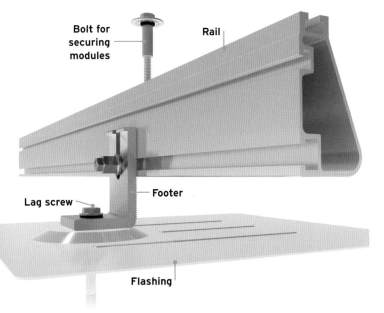

Bolt for securing modules

Rail

Lag screw

Footer

Flashing

Footers, flashing, and screws must be sealed with roofing sealant to prevent leaks.

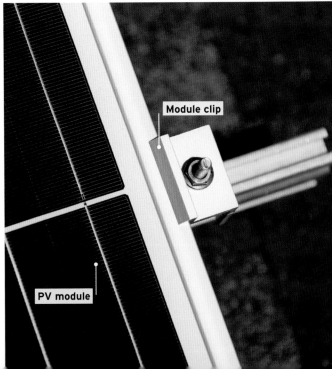

Module clip

PV module

Rails vary in size and strength to suit different climates and a range of snow and wind conditions. A midweight rail is suitable for most climates, but again, this is subject to local code. It's okay to use a heavier-duty rail than required for your area, but this adds unnecessary cost. Using an undersized rail is out of the question. Rails are typically available in 10-foot and longer lengths and are fastened together with special metal fittings called **splices** to create long continuous rows.

Another type of racking system is the **railless** style, which includes special footers but no rails. With some systems, modules are secured to a footer at each corner. With others, the modules are linked together with special brackets or splicing devices that are bolted onto the module frames; the joined modules form a rigid unit that spans across the footers.

On the upside, railless systems use less material than conventional rail-style racking, reducing shipping and overall system costs. Manufacturers often claim that railless systems install more quickly, too, but this benefit comes with experience, and the learning curve on railless racking can be relatively steep. Railless systems can also have a somewhat sleeker look, since there's no rail visible at the bottom or ends of the array.

On the down side, railless racking is less universal than rail systems. In some cases, modules need special framing to work with railless racking. Installing a railless system for the first time may be tricky, adding time to the project. Installation also requires leveling the individual footers instead of leveling a few long rails. Finally, while railless systems may someday be the standard, they're still the relative newcomer in the industry. Module warranties, code requirements, and design specs typically assume rail-mount systems, so if you go railless make sure your system is fully compliant and covered, and allowed by your local building authority.

TIP

CRITTER-PROOF YOUR ARRAY

The sheltered space between the roof and your PV modules can become a peanut gallery for squirrels and other rooftop ne'er-do-wells. Animals have been known to chew through wires, causing electrical failures. Keep them out with a simple screening system available from racking manufacturers. The screen material can be trimmed to fit your space and installs with a simple clip system.

Screen systems also are used for ground-mount arrays and are required for some installations. In such cases, a flexible screen extending from the ground to the modules can be installed after the array assembly is complete.

Roof mounts and skirting installed for a railless mounting system. The skirting is optional and provides a finished look to the most visible portion of the array.

DON'T HAVE ASPHALT SHINGLES?

Standard rooftop racking footers are made for asphalt (composite) shingle roofs, since that's what most houses have. Slate, tile (concrete or clay), and wood shingle or shake roofs require specialty anchors that serve as footers for conventional racking systems. These add cost and some installation time, but once the anchors are in place the rest of the PV installation is the same as with an asphalt shingle roof. Just make sure the anchors you choose are compatible with your racking system.

Also check with the anchor/footer manufacturer for installation details. Wood roofing is relatively easy for installing arrays, while slate and tile roofs are more challenging, and you can cause a lot of damage just being up on the roof, let alone cutting into it. If your roof is made from specialty material, it's a good idea to consult with a roofer or solar professional who has experience with that material.

If you're the proud owner of a healthy standing-seam metal roof, you're in luck. You can use specialty mounts that clip onto the seams — no penetration required! These specialty mounts, called S-5 clips, are fastened to the standing seams, and the rails and/or modules are then attached to the clips.

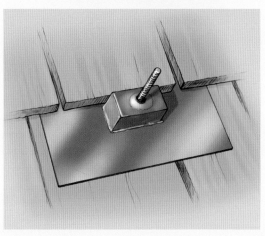

Cedar shake/shingle roof with flashing and footer base

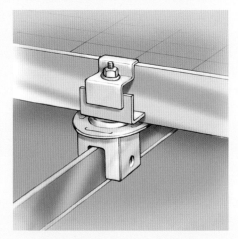

Standing-seam metal roofing with S-5 clip supporting module/rail

Grounding Your Modules and Racking

The racking and all the modules of a PV array must be electrically grounded (bonded) and connected to a ground wire leading to a system ground that terminates either in the home's main electrical service panel or onto a ground rod buried in the earth, depending on the system and local building code requirements. In the not-so-distant past, modules and racking rails had to be bonded with a long bare-copper ground wire that was attached to each module and each rail as it snaked its way from one end of the array to the other. That all changed several years ago with the advent of the **Washer Electrical Equipment Bond (WEEB)**.

The WEEB is a flat metal washer that installs between the rail of the racking and the module frame and is secured by the module's mid clamp or end clamp. Little teeth on the bottom side of the WEEB bite into the rail when the clamp is tightened, piercing the factory-applied anodized coating (as well as the natural layer of aluminum oxide that forms on all aluminum) to create a reliable bond. The WEEB is UL-listed, so it is now the universally accepted grounding device for PV arrays.

Module mid clamp with teeth for bonding the module to the rail (no separate WEEB needed). Mid clamps bond two module frames at once. The teeth face down to bite into the metal frames.

Each pair of modules requires two WEEBs total. If you have 10 modules on a row of racking, that's 10 WEEBs. But if you have an odd number of modules, you need two more WEEBs for the last module at the end of the row; for example, a row of 11 modules requires 12 WEEBs (you will count how many you need in the design phase). With the modules bonded to the rails using WEEBs, all you have to do is bond all the rails with a bare copper ground wire. Some racking manufacturers have taken the WEEB idea a step further by integrating metal teeth into their module clamps so you don't need separate WEEBs for those parts.

NOTE: *Some microinverter systems may need a continuous ground wire to bond all of the microinverters together. This can be the same wire that grounds the rails (using grounding lugs attached to the rails). Microinverter systems designed with integrated grounding may not need an external ground wire for the microinverters, but the racking still needs its own ground.*

WEEBs come in a variety of shapes and sizes to work with different racking systems and module and clamp configurations.

Mounting Structures for Flat Roofs

Weighted, or ballasted, systems are the standard racking option for houses with flat roofs. There are many different types available. Some consist of angled brackets or bars that are bolted together to support the modules and connect rows to one another (see pages 124–125). Other types have horizontal base rails extending from row to row, with tilt legs that support the modules at the designed tilt. The latter type often can accommodate greater tilt angles than the former.

Most systems are site-assembled and weighted with concrete blocks (the solid, 1- to 2-inch-thick type, not the big blocks with holes for building walls). Installation is simple and straightforward, and because you're not anchoring through the roofing material, you save all the steps and worry of sealing penetrations against leaks.

Ballasted racking is limited to relatively low-tilt angles, typically a maximum of 10 to 20 degrees. If greater tilt is critical, designers may opt for additional ballast blocks or an anchored racking system that mounts on stanchions. These are used most commonly in commercial applications and where a significant module tilt is required. Stanchions may also be used on "flat" roofs that have a slight slope (perhaps 10 to 15 degrees). Anchoring offers the high level of wind resistance required for steeply tilted modules but makes for a much more complicated installation, not to mention the risk of leaks at the penetrations. Anchored systems may require professional design and engineering.

Flat-roof system with horizontal base rails weighted with concrete blocks (ballast)

Most ballasted systems can accommodate any standard PV module, but of course you should confirm this up front. Design and installation of ballast systems are product-specific. Refer to the manufacturer for installation details and system specs. A ballasted array typically has a landscape layout because of its low profile (see Module Orientation: Portrait or Landscape?, page 49). Anchored systems can be landscape or portrait orientation.

TILTED RACKING

Tilted racking is sometimes used on low-slope roofs, flat roofs, and even opposing slopes (to create a south-facing array on a north-facing slope, for example). The tilt is made possible with fixed or adjustable tilted racking or with "tilt legs" added to standard flush-mount footers.

The idea behind tilting is to put the modules at a more favorable angle to increase production, but on most sloped roofs the difference is not significant. You can easily test this with PVWatts by comparing results with your roof's actual tilt against results with the adjusted tilt using a tilted rack. As mentioned in chapter 2, the gains you achieve in one season (summer, for example) may be offset by losses in the opposite season (winter), resulting in only a small net gain in annual production.

Tilted racking may face special restrictions from building authorities (tilting increases wind loads on the modules, which together act like a large sail), zoning boards, and homeowner's associations (a tilted array has a less integrated look than a flush-mount structure). Adjustable racking allows the owner to change the tilt with the seasons, but experience suggests that most people with this option don't bother with it after the first year or two.

Ground-Mount Structures

Manufactured ground-mount structures may have a dauntingly industrial look, but many have elegant, simple designs that are surprisingly well suited to DIY installation. Almost all residential systems consist of a corrosion-resistant metal framework anchored to the ground with concrete. The main ground-mount structure supports standard module rails, so the module installation is similar to that of a rooftop array, including the grounding system (see Grounding Your Modules and Racking, page 43).

There are two main types of ground-mount structures: standard ground-mounts and pole-mounts.

STANDARD GROUND-MOUNTS are typically fixed-tilt (nonadjustable) arrays that can have a couple of different types of construction. The most common construction consists of two rows of vertical posts — one row of short posts in front and one row of long posts in back, to create the tilt. The posts, usually pipes or manufactured columns buried in concrete, support the framework that holds the array. The other type of construction is a cantilever design, with angled supports that hang off a single row of vertical posts.

TIP

CAN I BUILD MY OWN WOODEN RACK?

You've probably seen DIY wooden ground-mount structures online or elsewhere. This might be allowed in your area, but it likely requires an engineer's stamp; that adds time and cost to the design. And lumber isn't cheap, especially good, stout, pressure-treated lumber that needs to last a few decades outdoors. Then there are framing connectors, fasteners, and concrete. And that's just the structure; you still need manufactured rails and hardware for the array. When you add up the cost and trouble of a homebuilt rack, a prefab metal ground-mount system starts to look pretty good.

On large commercial projects, the vertical supports may be piles (metal beams or poles) or earth screws (like giant wood screws) driven into the ground with heavy equipment. If you're planning a very large residential system — say, 20 kW or more — a system with driven piles is worth looking into, but otherwise it is not cost-effective. Therefore, most residential systems use concrete piers.

POLE MOUNTS fall into two subcategories: single-pole and multi-pole structures. **Single-pole mounts** include a single vertical pole that supports an array framework mounted to the top of the pole (called top-of-pole) or to the side of the pole (side-of-pole). Side-of-pole mounts typically are used for small arrays of one to four modules and may offer some tilt adjustment. Top-of-pole mounts typically handle about 10 to 12 modules (large systems can support upward of 20) and are highly adjustable. This is the type to use if you want to include a tracking motor.

MULTI-POLE MOUNTS have one or more rows of vertical support poles with top-of-pole framework joining the vertical supports. This may be the best option if you want some adjustability but won't use a tracker, and you'd like to have a sizable array on a single mount; multiple poles don't have to be as large or buried as deeply as single poles. Most systems offer a range of tilt angles, and many can be adjusted from the ground by one person (although it's usually much easier with two).

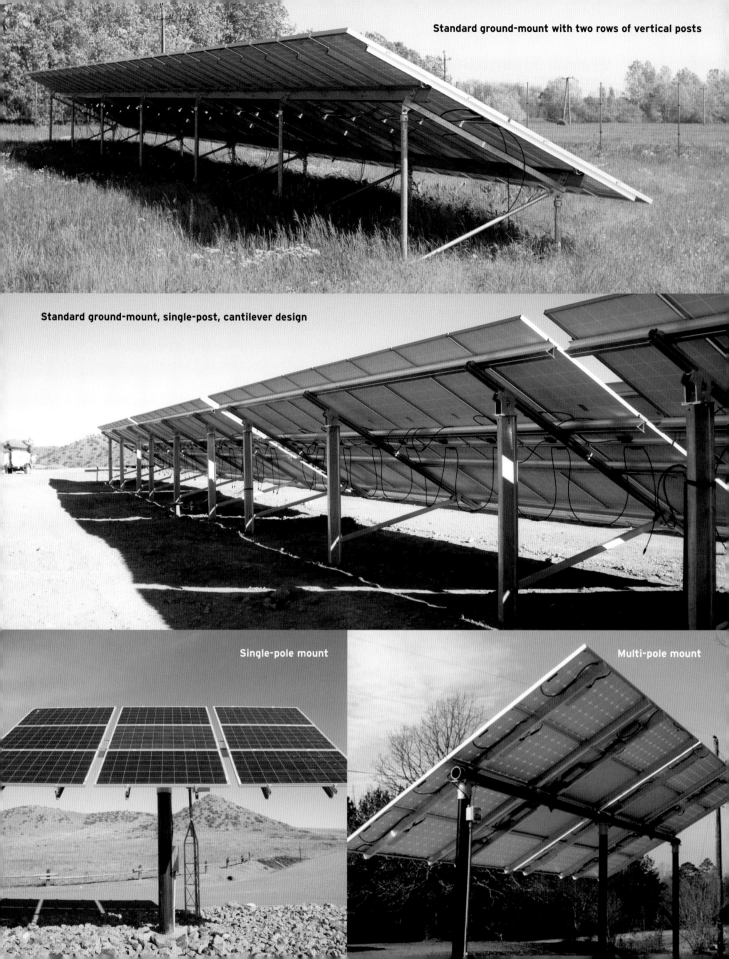

Standard ground-mount with two rows of vertical posts

Standard ground-mount, single-post, cantilever design

Single-pole mount

Multi-pole mount

Ground-Mount Design and Materials

Ground-mount support structures are usually custom-designed for each project. You give the manufacturer or solar professional your location (for wind, snow, soil conditions, and other climatic factors) and your system requirements (array size, tilt, orientation), and they design the racking system and include plan drawings and specs sheets for the installation. They may require a soil test from your site to complete the design. The manufacturer's drawings and specs might be sufficient for obtaining a project permit, or your building department may want to see a local engineer's stamp of approval. Before ordering a ground-mount system, learn about the building department's requirements so you can discuss them with the manufacturer.

Materials for standard ground-mount systems vary by manufacturer and model, but most structures start with vertical poles made of standard galvanized water pipe (typically schedule 40) that you can pick up at a local home center or plumbing supply store. Some systems use beams (H- or I-shaped) for posts; these are provided by the manufacturer or a solar equipment supplier. The base rails, which span across the vertical posts, may also be galvanized pipe, or they may be rails supplied by the manufacturer. The module rails span the base rails and support the modules. Rail placement, length, and quantity differ with module orientation (see Module Orientation: Portrait or Landscape?, opposite). Some ground-mount systems are designed to accommodate a variety of module rails from other manufacturers, and others come with their own rails.

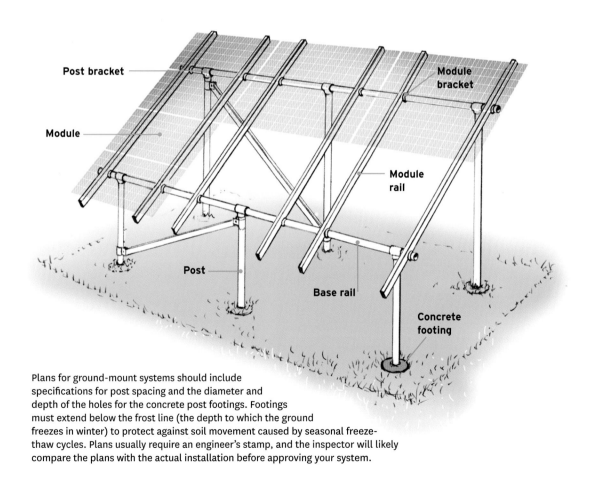

Post bracket

Module

Module bracket

Module rail

Post

Base rail

Concrete footing

Plans for ground-mount systems should include specifications for post spacing and the diameter and depth of the holes for the concrete post footings. Footings must extend below the frost line (the depth to which the ground freezes in winter) to protect against soil movement caused by seasonal freeze-thaw cycles. Plans usually require an engineer's stamp, and the inspector will likely compare the plans with the actual installation before approving your system.

MODULE ORIENTATION: PORTRAIT OR LANDSCAPE?

PV modules don't care whether they're vertical (portrait) or horizontal (landscape), so the choice is up to you. Usually it comes down to which way the panels fit best and what works well with the racking. However, the system design is specific to the orientation, so you do have to decide up front.

ON SLOPED ROOFTOPS, most arrays are portrait because this gives you more flexibility for the rail placement. With a portrait layout, the module rails are perpendicular to the rafters and can be set at any distance apart as long as you stay within the module manufacturer's specifications. With a landscape layout, the rails run parallel to the rafters and must follow the rafter spacing. Also, a portrait orientation on a sloped roof tends to look more natural. If you're considering a railless mounting system, however, be aware that some work only with a landscape orientation.

FLAT ROOFS typically call for landscape orientation to keep the top edges of the modules closer to the roof, reducing wind loads and the visibility of the modules. However, arrays can also be designed for portrait orientation, with appropriate engineering and design, usually done by professionals.

Portrait

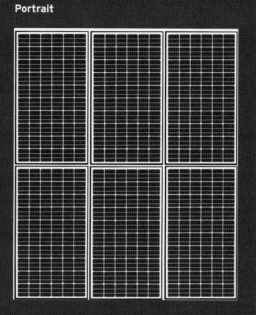

GROUND-MOUNT STRUCTURES can go either way, and orientation often is based on the number of modules and the racking design. A portrait layout usually has two horizontal rows of modules. A landscape layout can be two to five modules high.

Landscape

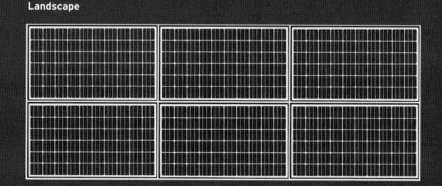

Inverters and DC Optimizers

All PV systems include one or more inverters that convert the solar-generated DC power to AC power for use in the house and for utility hookup for grid-tied systems (utility power also is AC). Options for grid-tied systems are discussed here. Off-grid systems have a somewhat different setup, which typically includes one or more inverters that receive DC power from the batteries and send AC power to the house. Off-grid inverters are discussed in chapter 9.

There are two types of inverters for grid-tied systems: string inverters and microinverters. A third option is to add DC optimizers to a string-inverter system.

STRING INVERTERS are relatively large units that mount (usually on a wall) at the ground-floor level, often near the hookup components and usually in a garage or basement. Most units have outdoor-rated enclosures and can also be mounted on the exterior wall. They receive DC power from the series-strings of the array and deliver AC power to the electrical loads of the house. Most residential PV systems with string inverters have one or two inverters, depending on the array size and layout. String inverters for grid-tied systems are commonly referred to as "grid-tie inverters," to distinguish them from *stand-alone inverters* (sometimes called "battery inverters") used for off-grid systems. This is standard industry terminology you'll see when shopping for inverters.

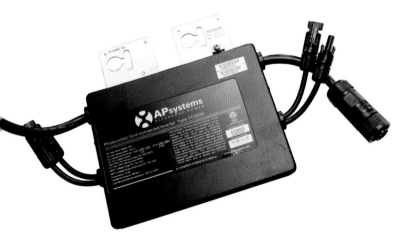

MICROINVERTERS are small units that mount onto the back side of each module or onto the racking system below the modules. Typically, each module has its own microinverter, although some microinverters can be fed from two modules. Microinverters convert DC power to AC power at the individual module, so all of the wiring that connects the modules and runs down to the ground carries AC electricity.

Performance Monitoring and Optimization

Inverters do more than convert power from DC to AC. They also monitor the modules' or strings' performance and automatically make adjustments to optimize power output. Solar electricity production constantly fluctuates throughout the day, due to changes in solar irradiance (how much sun is hitting the modules), air temperature, shading, and other factors. Inverters monitor the effects of these changes and respond by shifting the balance of voltage and amperage to get the highest possible output from the modules. This is known as **maximum power point tracking (MPPT)**, and all modern string inverters and microinverters have this capability. Inverters are also the tools with which PV system owners and technicians can monitor the output and other performance factors of their modules or module strings, aiding maintenance, troubleshooting, and service.

You'll get a better understanding of how monitoring and optimization work when you learn about I-V curves (page 61), but for now you just need to know that string inverters and microinverters handle performance monitoring differently. As mentioned briefly in the discussion about shade in chapter 2, string inverters get input from modules as strings, not as individual modules. The inverter monitors and optimizes the output of the entire string because it can't distinguish one module from the others.

Microinverters offer module-level monitoring and MPPT optimization. If a module is shaded, its microinverter can adjust its production without affecting the other modules. Module-level monitoring also gives you the flexibility to design the array with small groups of modules on different roof slopes and even to use different sizes and models of modules. Individual module monitoring also helps with troubleshooting problems. However, service of microinverters does require module removal.

DC optimizers mounted to rails

To help compensate for this inherent difference, there are two ways to give string inverter systems more precise monitoring. One is to use a string inverter with multiple inputs, each with MPPT capability. This allows different strings (not modules) to have different orientations, tilts, and module types without having one string hinder the performance of the others. The second option is to add **DC optimizers** to a string inverter system. Like microinverters, DC optimizers are installed at each module and make adjustments on an individual basis. The key difference between microinverters and DC optimizers is that the latter do not convert DC power to AC power at the module; they are always used on systems with a string inverter for the power conversion. They do, however, allow for individual module monitoring and MPPT optimization.

You can check on your modules' performance with a data monitoring system that you access on your computer or smartphone, through the manufacturer's online software (see Monitoring Your System, page 180). As you might guess, string inverters can tell you what's happening only with each string in your array, while microinverters and DC optimizers report on each module. If a module is significantly underperforming for no apparent reason, you'll know something is wrong with it. Same thing with a microinverter or optimizer that's gone offline.

Data monitoring capability is not standard with all string inverters, so be sure to consider whether you'd like to have this feature. With some inverters, it's an add-on that you have to order. When it's time to set up the monitoring system after installation, you can contact the manufacturer to have a member of the support staff walk you through the process.

DC optimizer

STRING INVERTERS VERSUS MICROINVERTERS

There's no denying that microinverters (the young upstarts) have some nifty features and may seem like a no-brainer for your new PV system, but there are some good reasons why solar pros still opt for string inverters (the old stalwarts) when designing for residential customers. Here's a quick break-down of the basic pros and cons of each.

MICROINVERTERS

PROS

- Microinverters offer module-level performance monitoring and optimization.

- Design flexibility means fewer restrictions when arranging modules (explained in chapter 4).

- Rapid-shutdown equipment is not required, which saves installation time and cost.

- Failure/problem with an individual microinverter affects only one module, not the entire system.

- AC output is at the module level.

CONS

- The total cost of all microinverters may be higher than the cost of a single string inverter (even after factoring in savings on rapid shutdown equipment).

- Sometimes the microinverter output becomes restricted under high irradiance (for example, the module may be producing 300 watts but the microvinverter output is limited to 280 watts).

- Replacement requires removal of one or more modules; Murphy's law dictates that the bad microinverter will be buried deep in the array, so you would have to take off multiple modules to reach it.

- Some PV owners and installers, fearing wear and tear, don't like the idea of having sensitive electronics on the roof, especially when the effects of extreme temperatures are a risk.

- Microinverters cannot be used with standard off-grid systems.

STRING INVERTERS

PROS

- String inverters install at ground level, often in a protected area, offering easy access and a single point of maintenance/service.

- They can be less expensive, particularly with large systems.

- They can have an auxiliary/backup power supply (available only on select units; see page 55).

CONS

- A string inverter failure or problem disables the entire PV system.

- Performance monitoring and optimization are typically less precise — at the string level, not the module level — though this varies by system and inverter.

- A string inverter system requires rapid-shutdown equipment, which means additional cost, installation time, and potential ongoing maintenance or future replacement.

Rapid Shutdown Protection

One of the big benefits of microinverters bears explanation. Under the National Electrical Code (NEC), PV systems are now required to have **rapid shutdown** protection. This means power between the solar array and the DC-AC inverter can be shut down both automatically and manually.

Here's why rapid shutdown is needed: When the grid goes down, inverters automatically cut power to the wiring and components **between the inverter and the main electrical service panel** (breaker box). This "self-islanding" feature prevents the PV system from backfeeding the grid during an outage or when the utility has shut off the power to work on the lines. When the grid is not down, the DC disconnect switch (included with all new string inverters) provides a manual means to cut the solar-generated power between the inverter and the service panel.

However, PV modules always produce electricity during daylight hours. That means they're feeding the wires **between the array and the inverter** even when the power is shut off at the DC disconnect. Rapid shutdown cuts the power to this wiring so that both utility workers and firefighters and other emergency responders are not exposed to a shock risk if they're working around the wiring or other components on the **DC side** of the system.

Microinverters come with automatic shutdown capability. PV systems with string inverters require a special rapid shutdown system for code compliance. This may include an external box with a manual switch linked to a special disconnecting combiner box at the PV array; alternatively, the system may be designed to use DC optimizers to provide this automatic shutdown feature, which may require special connectors or an add-on device. The additional equipment adds cost (and some installation time) to a string-inverter system.

Rapid shutdown emergency disconnect switch unit

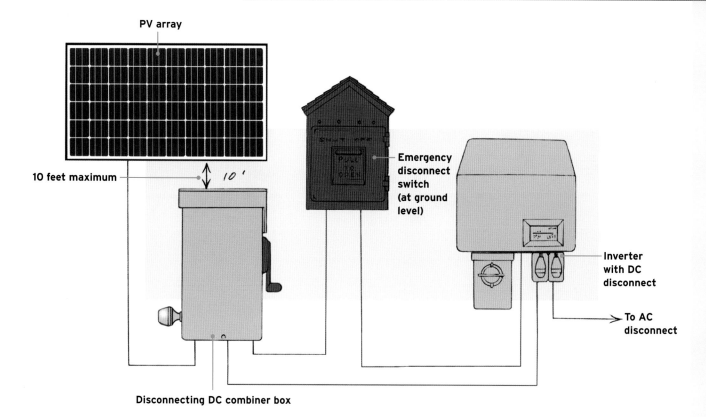

PV array

10 feet maximum

10'

Emergency disconnect switch (at ground level)

Inverter with DC disconnect

To AC disconnect

Disconnecting DC combiner box

A standard rapid shutdown system includes an emergency disconnect switch at ground level and a disconnecting combiner box located within 10 feet of the PV array. Pushing the switch on the emergency disconnect causes the combiner box to shut down the power between itself and the inverter. The wiring between the array and combiner box remains live, which is why the box must be no more than 10 feet from the array.

EMERGENCY POWER FOR GRID-TIED SYSTEMS

When the grid goes down, your grid-tied system goes down due to the self-islanding feature, but the array still produces power during daylight hours. A special feature on some grid-tied string inverters allows you to tap into this power and feed it to a single electrical outlet when the grid goes down. The outlet typically supplies a maximum of 1,500 watts and works only during the daytime when the array is productive, but that's plenty of power for charging electronics or running lights or small appliances. This capability is not available for microinverter systems.

INVERTERS AND DC OPTIMIZERS

55

4 Designing Your System

Create your entire PV system on paper

THE DESIGN PROCESS BEGINS with narrowing down your options for modules and other hardware. It ends with a simple but detailed array layout (with specifications) and a wiring diagram of the entire PV system. You'll use these when purchasing the system hardware and when applying for a permit. Along the way, you'll learn some basic principles of electricity and make some simple calculations (no difficult math, we promise). This is the most technical part of a solar project, and it can be a lot to digest at first. (If your mind were a PV system, it would be best to save the design work for a sunny day.) But once you absorb the basics, the confidence you've gained will stay with you for the rest of the project.

Note to off-grid folks: The general focus and the sample design in this chapter involve grid-tied systems, but all of the design information pertains to off-grid systems, too. Design information specific to off-grid systems is covered in chapter 9.

PV Circuit Fundamentals

Before designing your system, you need to understand a few principles of electricity: (1) series and parallel wiring, (2) a set of electrical circuit rules, and (3) I-V curves. We'll discuss the basic science of each and leave you with some very simple takeaways to apply to your own calculations. These basic principles are important whether you're designing for string inverters or microinverters.

Series and Parallel Wiring

Series wiring and parallel wiring are two configurations used for different parts of a PV system. Individual modules are wired together in series to create the series-strings. The series-strings are then wired together in parallel before feeding the rest of the system. This applies to grid-tied systems with string inverters and to off-grid systems. Off-grid folks also will apply series and parallel wiring to the battery bank (see page 160). With microinverter systems, you don't have to bother with parallel wiring, as explained on page 60.

Wiring modules in series means that you join the positive (+) lead of one module to the negative (-) lead of the next. At the end of this string, you have one extra negative lead on one end and one extra positive lead on the other end.

The effect of series wiring is that the **voltage adds up** while the **amperage stays the same**. For example, if you have 10 modules in a series string, and each has an output rating of 30 volts at 8 amps, the string will have a voltage rating of 300 volts (10 × 30) at 8 amps of current.

Wiring series-strings of modules in parallel means that you connect the extra positive leads of the strings together and the extra negative leads of the strings together. The effect of parallel wiring is that the **voltage stays the same** while the **amperage adds up**. For example, if you have two of the above strings (300 volts at 8 amps) and wire them in parallel, the entire circuit will have a rated output of 300 volts at 16 amps. (Remember that volts × amps = watts; therefore, this array has a rated power output of 300 × 16 = 4,800 watts.)

The reason that you wire modules in series and you wire series-strings in parallel — and not the other way around — comes down to equipment capability and compatibility. PV modules output a relatively low voltage (typically about 30 volts DC) and a relatively high current (typically about 8 amps DC). Compare that to a standard household electrical circuit, which operates at 120 volts AC and a maximum of 15 or 20 amps AC. Wiring the modules in series brings up the voltage without raising the current. Also, string inverters typically are designed to accept 200 to 400 volts and 10 to 40 amps. Wiring modules in series and series-strings in parallel allows the PV array to hit those ranges for voltage and current.

DC Electrical Circuit Rules

Three electrical circuit rules (ECRs) govern the behavior of basic DC electrical circuits wired in series and parallel. These are based on scientific principles known as Kirchoff's laws. (You already know the gist of #1 and #3.)

ECR #1: When modules are wired in series, the voltages add up while the amperage stays the same.

ECR #2: When series-strings of modules are wired in parallel, the voltages of each of the strings are forced to be the same.

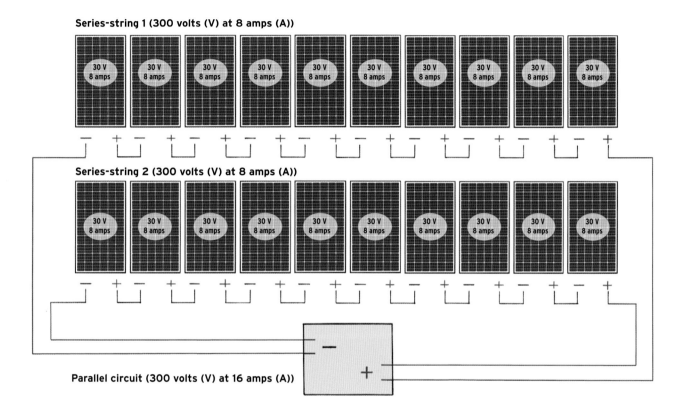

Series-string 1 (300 volts (V) at 8 amps (A))

30 V 8 amps · 30 V 8 amps · 30 V 8 amps · 30 V 8 amps · 30 V 8 amps · 30 V 8 amps · 30 V 8 amps · 30 V 8 amps · 30 V 8 amps · 30 V 8 amps

Series-string 2 (300 volts (V) at 8 amps (A))

30 V 8 amps · 30 V 8 amps · 30 V 8 amps · 30 V 8 amps · 30 V 8 amps · 30 V 8 amps · 30 V 8 amps · 30 V 8 amps · 30 V 8 amps · 30 V 8 amps

Parallel circuit (300 volts (V) at 16 amps (A))

Each of these two series-strings shows modules wired together in series (positive to negative leads). The two series-strings are then wired together in parallel (positive to positive leads and negative to negative leads).

ECR #3: When series-strings of modules are wired in parallel, the total amperage of the strings adds up while the voltage of each string stays the same.

You will apply these three rules when calculating the output values for your modules and strings and when making sure the values are compatible with your inverter's input requirements and specifications. They also affect how you design the module strings, but the rules apply differently to systems with string inverters and microinverters.

STRING INVERTERS: ECR #2 dictates that every series-string in an array should have the same number of modules. For example, if you've determined that your array will have 18 modules total, you can wire them into two strings of 9 modules or even three strings of 6 modules, but you should not wire them with one string of 10 and one string of 8. Since ECR #2 forces both strings of the parallel circuit to have the same voltage, if you have uneven strings one will be operating above its optimal value and one will be operating below its optimal

level. Therefore, neither string will be operating at its optimal voltage point. You'll understand this more fully with the following discussion of I-V curves.

MICROINVERTERS: Since microinverters convert from DC to AC at the module, the above rules don't apply. This means you don't have to worry about grouping all of your modules in equal numbers. For example, it's okay to have one string of 10 modules and one string of 8, or even two strings of 7 and one string of 4, without compromising performance. What you do have to follow is the manufacturer's limit on how many microinverters can be connected together. Some microinverters can be joined in large strings of 20 or more modules, so some arrays may be made up of a single string. But most microinverters have lower limits, resulting in arrays consisting of two or more strings. By the way, module groupings in microinverter systems are called **branch circuits**, perhaps to avoid confusion with string-inverter strings. Both terms refer to a group of modules wired together in series.

PV CIRCUIT FUNDAMENTALS: TAKEAWAYS

- When you wire modules together in *series* you add up the *voltage* of all the modules in the series-string to determine the string's output voltage (ECR #1).

- The voltages of all of the series-strings in a *parallel* circuit are forced to be the same (ECR #2). Therefore, arrays for string inverter systems must have the same number of modules in each string.

- When you wire series-strings together in *parallel* you add up the *amperage* of all the strings to determine the circuit's output current (ECR #3).

- Strings (branch circuits) in microinverter systems must not exceed the microinverter limit of maximum units per string.

I-V Curve

In chapter 3, in the section Performance Monitoring and Optimization (page 51), we discussed how inverters use maximum power point tracking (MPPT) to adjust voltage and current for the best possible power output under the present set of outdoor conditions. The I-V curve helps demonstrate the basic science behind the MPPT feature on inverters and helps you understand the module specs you will use when designing your system.

An I-V curve is a graphic representation of the relationship between the current, or amperage (I), and voltage (V) in an electrical device such as a solar cell or PV module. Since electrical power (P, measured in watts) is a product of current and voltage, an I-V curve shows how changes in either the current or the voltage affect electrical output:

$$P \text{ (power)} = I \text{ (current)} \times V \text{ (voltage)}$$
$$\text{and}$$
$$1 \text{ watt} = 1 \text{ amp} \times 1 \text{ volt}$$

As you can see in the I-V curve of a single PV module (below), when the voltage is at its highest point (40 volts in this example) the current is at 0. That means 0 electricity (40 volts × 0 amps = 0 watts). Conversely, when the current is at its highest point (9 amps) but the voltage is at 0, again there is no power (9 amps × 0 volts = 0 watts). This tells you that in order to create usable electricity, you need to strike a balance between the current and the voltage.

A brief background lesson: Voltage is loosely described as the *pressure* in an electrical circuit that forces the electricity to flow. Think of water pressure in a garden hose. Current is described as the *flow* of electrons in a circuit and is a result of the pressure. Think of the volume of water flowing in a garden hose. If you place a PV module in full sunlight and connect its plus (+) and minus (–) output wires together, you create a short circuit with maximum current (high flow) but zero voltage (no pressure) because there's no resistance to the flow. This is known

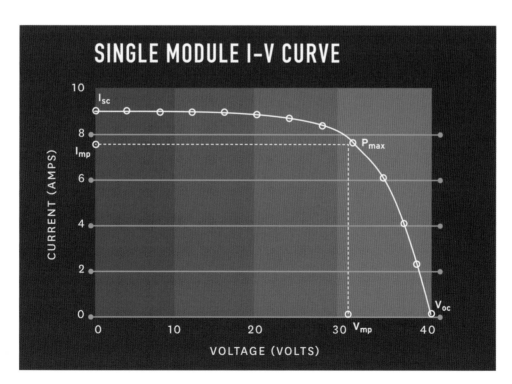

SINGLE MODULE I-V CURVE

as the **short-circuit current** value **(I$_{sc}$)** on the module's spec sheet and can be seen on the I-V curve as the point where the curve intersects the Y (current) axis. If you disconnect the two wires, you create an open-circuit condition that has maximum voltage (high pressure) but zero current (no flow). The solar cells are active, but there's nowhere for the electrons to go, so you get no electricity. This is the **open-circuit voltage** value **(V$_{oc}$)** on the module's spec sheet and is the point on the I-V graph where the curve intersects the X (voltage) axis.

Back to the I-V curve: At every point along this curve there is a power value (in watts) associated with that point's current (amps) times its voltage (volts). You get the most electricity when you have the highest combined values of both current and voltage along the I-V curve; that is, when the I × V product is the greatest. This sweet spot is called the **maximum power point (MPP, or P$_{max}$)**. This is always near the "knee" of the I-V curve; in this example, P$_{max}$

is at about 33 volts and 7.6 amps. This is also where we want the module to operate and is the same as the module's rated output (33 volts × 7.6 amps = 250.8 watts).

DC-AC inverters (both string inverters and microinverters) detect current and voltage in a module's or string's circuit and then change the level of resistance in the circuit to adjust the voltage and current in order to hit the MPP at all times. If the current goes down (because the sun went behind a cloud, for instance), the inverter adjusts resistance to change the voltage and maintain the MPP at the lower current value. When the sun comes back out and the current goes up, the inverter again adjusts the resistance, changing the voltage and increasing the output, again to operate at the MPP. This process of continual MPPT optimization provides the homeowner with the maximum electrical output of their PV system at all times, no matter what the outdoor weather conditions are.

The Design Process

Now that you know the fundamental rules of PV circuits, it's time to begin the design process in earnest. PV system design largely revolves around the three main components we looked at in chapter 3: the modules, the module support structure, and the DC-AC inverter(s). The process follows a sequence of steps that starts with your DC system size and the AC annual energy goal you calculated with PVWatts in chapter 2.

With your goal in mind, you want to choose a module that's a good candidate for reaching this goal and that also fits on your roof and works with your budget. Next, you use the module specs (readily available online) to plan the array layout and wiring configurations (strings) and run the calculations for the inverter(s). Your choice of rooftop racking or ground-mount structure goes along with the module model, since they must be compatible. Once you have the three main pieces (modules, support structure, inverters) in place, the rest of the components are easy to fill in, as they're largely generic. The basic process is outlined in the flowchart on page 64.

There are a couple of rules to keep in mind as you work through this process:

1. **BE FLEXIBLE.** Because all of the system components must be compatible, and together they must meet your power goal (and budget), you might end up swapping products here and there before finalizing your shopping list. Don't set your heart on one model of module and force everything else to fit. There are many ways to achieve the same goal, and tweaking the system design as you go is a normal part of the process.

2. **YOUR FINAL SYSTEM DESIGN MUST WORK BOTH PHYSICALLY AND ELECTRICALLY.** Finding modules that will fit on your roof without violating installation restrictions (and also look good, if that matters) is an example of a physical requirement. Finding modules with electrical specs that meet your power goals and work well with your inverter is one of the electrical requirements. An additional electrical requirement is to make sure your series and parallel module strings do not violate ECRs #1-#3 (see page 58).

PV SYSTEM DESIGN FLOW CHART

GOALS
- DC system size (in kW) from PVWatts, based on AC annual energy goal (in kWh/year)

MODULE SELECTION
Check module specs for:
- Output (wattage, voltage, current)
- Module dimensions
- Compatibility with support system

ARRAY LAYOUT
- Determine total number of modules and plan array physical layout on roof/ground.

ELECTRICAL CALCULATIONS
- Apply module specs to calculate string and system voltage, current, and wattage.
- Confirm array layout and string configuration.

INVERTER SELECTION
- Apply inverter specs to electrical calculations. Make sure inverter is compatible with module strings (series and parallel).

MECHANICAL CALCULATIONS
- Determine quantities of racking/ground-mount elements (footers, rails, posts, clamps, etc.).

METERS, DISCONNECTS, AND SO ON
- Note remaining components in order of installation location. (Your electrician can help with this.)

Sample System Designs

The best way to learn system design is to study some realistic examples. We'll start with walking step-by-step through Sample Design 1: a standard sloped-rooftop (flush-mount) system with a string inverter (including variations for DC optimizers and flat roofs). No matter what your plans are, you should read through this sample process because it covers the basic design process in detail. Then we'll look at a rooftop system with microinverters (Sample Design 2) as well as two ground-mount systems (Sample Designs 3 and 4), focusing on the points where they differ from Sample Design 1.

The sample worksheets shown here are based on actual designs for PV systems that now power actual households. Using a worksheet template like this is a nice way to consolidate all of the design essentials on one piece of paper. It will serve as a handy reference when you are filling out permit forms and making your hardware purchases. You can create a similar template using a spreadsheet application such as Excel and modify it as needed.

Rooftop Array with String Inverter

This sample system is a 4.5 kW grid-tied system with a single string inverter. The solar array contains 18 modules laid out in two physical rows of 9 modules each and two series-strings of 9 modules each. Note that in rooftop arrays, the physical rows don't have to be laid out the same as the electrical strings; for example, you could have three rows of 6 modules each and wire them as two series-strings of 9 modules each. The rooftop area is roughly 32 × 14 feet, which leaves plenty of room for the array and the installation margins typically required by the building department. On the following pages are steps for completing the design and filling out each section of the template; the sample values are given in **blue**:

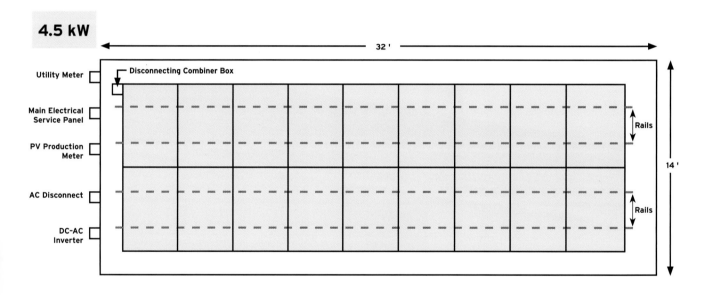

PV Array Layout: 2 rows of 9 modules each; portrait orientation

DC–AC INVERTER SPECS

Manuf.: Fronius

Model: IG Plus 5.0

MPPT Voltage Range: 230–500 Volts

Max. System Voltage: 600 Volts

Max. DC Input Current: 23.4 Amps

AC Output Voltage: 208/240/277 Volts

PV MODULE SPECS

Model: Helios 6T-250

P_{max} = 250 Watts

V_{pmax} = 30.3 Volts

V_{oc} = 37.4 Volts

I_{pmax} = 8.22 Amps

I_{sc} = 8.72 Amps

Length = 66.1 Inches

Width = 39.0 Inches

MECHANICAL SPECS

4 feet	Footer Spacing
355	Rail Length (inches)
4	# of Rails
36	# of Footers
8	# of End Clamps
32	# of Mid Clamps
20	# of WEEBs

SYSTEM ELECTRICAL SPECS

Module Orientation: Portrait Azimuth: 205° Tilt: 23°

18 x-Si PV Modules × 250 Watts = 4,500 Watts (DC, peak sunlight)

9 Modules per Series-String => 272.7 Volts @ 8.22 Amps = 2,242 Watts / String
(String Voc × 1.25 = 421 Volts)   (String Ampacity = 13.63 Amps)

2 Parallel Strings per 5.0kW Inverter => 272.7 Volts @ 16.44 Amps = 4,483 Watts / Inverter
1 × 5.0kW DC–AC Inverter => 4,483 Watts (DC, peak sunlight)   PVWatts: 6,294 kWh (AC/year)

TOTAL SYSTEM OUTPUT (PVWatts)

4,483 Watts (DC, peak sunlight)

6,294 kWh (AC, annual)

1. SITE CHARACTERISTICS

Draw a simple map of the rooftop with dimensions of the PV installation area. Include the ridge, eave, and any roof edges close to the array, as well as any penetrations or obstructions you will need to avoid during installation. Note the module orientation and the roof azimuth and tilt.

Module Orientation: **portrait**

Azimuth: **205** degrees

Tilt: **23** degrees

2. MODULE SPECS

The chosen module is the Helios 6T-250. Create a section for the module specs, noting the following values from the manufacturer's spec sheet:

P_{max}: **250** watts

V_{pmax}: **30.3** volts

V_{oc}: **37.4** volts

I_{pmax}: **8.22** amps

I_{sc}: **8.72** amps

Length: **66.1** inches

Width: **39** inches

3. NUMBER OF MODULES/ SYSTEM RATING

Divide the DC system size (4.5 kW) by the individual module wattage to find the number of modules in the array.

4,500 watts ÷ **250** watts per module = **18** modules

NOTE: The DC system size is the array size you calculated using PVWatts. The 4.5 kW (DC) system in this sample can produce about 6,294 kWh of AC energy per year, according to PVWatts. When doing your own design, confirm that the DC system size you're designing will meet your annual goal of AC energy production at your location, with your array orientation, azimuth, and tilt.

4. PHYSICAL ARRAY LAYOUT

In this example, the modules are in portrait orientation (yours may be portrait or landscape). To convert the side-to-side (east-to-west, in this example) roof dimension to inches, multiply by 12:

32 feet × 12 = **384** inches

Divide the total length by the module width:

384 inches ÷ **39** inches = **9.85**

Therefore, you can fit 9 modules in a row and have about 16.5 inches of space left on each side of the row:

9 modules × **39** inches = **351** inches

384 - **351** = **33** inches

33 ÷ 2 = **16.5** inches on each side

This is plenty of space to meet the 5 to 10 inches of spacing required by the building department (for wind uplift at the roof edges). Next, check the ridge-to-eave dimension to determine how many rows will fit on the roof:

14 feet × 12 = **168** inches

168 inches ÷ **66.1** inches (module length) = **2.54**

You can fit two rows of modules and have about 36 inches of space left over:

2 rows × **66.1** inches = **132.2** inches

168 - **132** = **36** inches

This leftover space could be divided equally, so you'd have 18 inches at the ridge and eave. But there are two good reasons to move the array toward the ridge: (1) more roof space below the array helps break up and contain snow that slides off the array, and (2) more space along the eave gives you a place to stand for the installation and future maintenance. In this case, a good trade-off is to leave 12 inches of space at the ridge and 24 inches along the eave. These both exceed the 5-10-inch requirement mentioned previously.

5. SERIES-STRING SIZE

Determine string size based on the total number of modules (divide into two, three, or more strings?) and installation area (which arrangement fits best on the roof, while meeting code requirements?). (See How to Measure Area, page 22.) Remember that series-strings must have an *equal number* of modules in each string for best performance.

18 modules total

9 modules per series-string

2 series-strings wired in parallel

If the total number of modules doesn't break up nicely into equal-sized strings, either add or subtract a module or two to the total, or choose a module model with a higher wattage rating so you can hit your power goal with fewer modules.

6. STRING WATTAGE/ INVERTER WATTAGE

Multiply the number of modules in each series-string by the module operating voltage (V_{pmax}), then by the module amps (I_{pmax}) to find the watts per string:

9 (modules) × **30.3** volts (V_{pmax}) = **272.7** volts

272.7 volts × **8.22** amps (I_{pmax}) = **2,242** watts/ series-string

Confirm that the wattage for the entire array is close to your system goal (4.5 kW):

2 parallel strings @ **2,242** watts each = **4,484** watts

7. MAXIMUM STRING VOLTAGE (extreme voltage calculation)

Multiply the number of modules in the series-string by the module's open-circuit voltage (V_{oc}), and then multiply by 1.25 (extreme temperature limit; this value is specified by the National Electrical Code, or NEC; use for all PV systems):

9 (modules) × **37.4** volts (V_{oc}) × 1.25 = **421** volts

8. STRING AMPACITY (extreme current calculation)

Multiply the module's short-circuit current (I_{sc}) by 1.25 (maximum irradiance value; use for all PV systems), and then again by 1.25 (NEC continuous-use value; use for all PV systems):

8.72 amps (I_{sc}) × 1.25 × 1.25 = **13.63** amps

9. INVERTER

Choose a DC–AC inverter based on the total calculated array wattage (4,484 watts or the DC system rating of 4,500 watts). Always select an inverter with a nameplate rating that exceeds your system's array wattage. The sample inverter is a Fronius IG Plus 5.0, a 5 kW inverter; 5,000 watts is more than the array wattage of 4,484. Follow this rule even when a manufacturer claims that an inverter is compatible with systems larger than the inverter's nameplate rating.

When you have chosen your inverter, create a section for inverter specs and enter the following values:

MPPT voltage range: **230–500** volts

Maximum system voltage: **600** volts

Maximum DC input current: **23.4** amps

AC output voltage: **208/240/277** volts

Using the values you calculated in the previous steps, confirm that the inverter specs work with your modules and string layout as follows:

MPPT VOLTAGE RANGE: 230–500 VOLTS.
The string operating voltage (**272.7** volts/string) must fall within this range, known as the *MPPT window.*

MAXIMUM SYSTEM VOLTAGE: 600 VOLTS.
Must be greater than the maximum string voltage (**421** volts). **Note:** 600 volts is the standard maximum system voltage for string inverters in the United States. Do not use an inverter with a maximum voltage of 1,000 volts (common in Europe), because many inspectors and utilities will not approve it. Further, if your maximum string voltage exceeds the inverter's 600-volt maximum, simply reduce your string size by one or two modules.

MAXIMUM DC INPUT CURRENT: 23.4 AMPS.
Must be greater than the total array current (amps): **2** series-strings at **8.22** amps (I_{pmax}) each, wired in parallel (amps add up) = **16.44** amps.

AC OUTPUT VOLTAGE: 208/240/277 VOLTS.
Most homes receive 240-volt power from the utility, but many commercial buildings use 208-volt or 277-volt power. You can confirm this with your utility. Make sure your DC–AC inverter is compatible with the appropriate voltage. Inverters with multiple voltage options have a separate input or switch for setting the desired voltage. Your electrician must set this, since it's on the AC side of the system.

10. MODULE SUPPORT STRUCTURE (racking system)

Work with the racking manufacturer to determine quantities of each part for the racking system, based on the module orientation and dimensions and the array size and layout. (If you have a flat roof, see pages 25 and 44.) In the sample design, the modules are in portrait orientation and have a standard racking system with rails. The quantities are calculated as follows:

 # OF RAILS: 4
2 rails per horizontal module row

RAIL LENGTH: 355 INCHES
9 modules per row @ 39 inches wide = 351 inches

Add 4 inches for spacing between modules (determined by module mid clamps and end clamps; 8 spaces @ ⅜ inch plus ½ inch for each end clamp): 351 + 4 = 355. This is the total length for each rail, which is made up of multiple rail pieces (typically 8 to 12 feet long) joined with splice fittings.

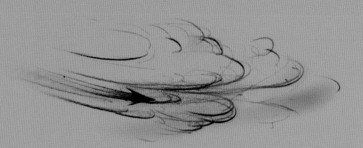

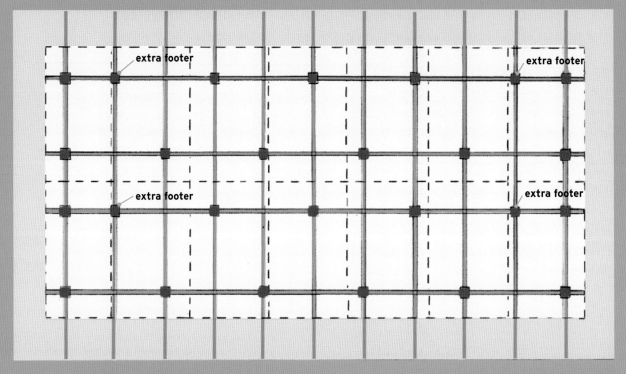

Within the diagram:
extra footer · extra footer
extra footer · extra footer

FOOTER SPACING FOR WIND AND SNOW

The outermost footers must be within 12 to 18 inches of the ends of the rail. For added protection against wind uplift on any exposed windward side of the roof (the side that usually receives the strongest winds), it's a good idea to place the next footer within 2 feet – wherever your next rafter falls – of the first. After that, space the footers at the normal interval (every 4 feet or as required).

If you live in an area with heavy snowfall, you can help distribute the load on your rafters by staggering the footers between rows. The example shown here assumes 2-foot rafter spacing. Note that the extra footer near the end applies to the top rail in each module row.

Sometimes an engineer may simply specify that one of the two rails in a row gets a footer on every rafter while the other rail in the row gets a footer every other rafter. In any case, more footers is better than fewer. You don't want your modules to blow off the roof or collapse under the weight of heavy snow loads.

OF FOOTERS: 36

Typical footer spacing is 48 inches. This can be confirmed by a professional engineer and/or the local building department. Start with end footers no more than 18 inches from each end of each rail, then space the footers every 4 feet, or as required by the racking manufacturer or the building department (see Footer Spacing for Wind and Snow, opposite, for additional considerations). Round up fractions to the nearest whole number. In this example, standard spacing is used, resulting in 9 footers for each of the 4 rails.

355 inches (rail length) ÷ **48** inches (footer spacing) = **7.4**

Round up to next integer: **8**

8 + **1** (for additional end footer) = **9** (footers per rail)

9 (footers) × **4** (rails) = **36** (total footers)

END CLAMPS: 8

Two end clamps for the modules at both ends of each row.

OF MID CLAMPS: 32

Two mid clamps between adjacent modules in each row. Clamps are not used between rows.

OF WEEBS: 20

Two WEEBs (see Grounding Your Modules and Racking, page 43) for every two modules. If the row has an even number of modules, the number of WEEBs will equal the number of modules in that row. If the row has an odd number of modules, add two more for the last module in each row.

11. DISCONNECTS AND METERS

Add all of the required disconnects and meters for the balance of the system. These must appear on the design in the same order in which they are wired and are usually mounted on the wall of the house at ground level (except the disconnecting combiner box). The order shown here is from the PV array to the utility-grid tie-in:

1. Rapid-shutdown disconnecting combiner box (typically on roof; must be within 10 feet of the array)

2. Rapid-shutdown emergency disconnect switch (such as Birdhouse)

NOTE: The rapid-shutdown switch installs between the combiner box and the inverter, but it is not part of the main circuit carrying DC power from the array.

3. DC–AC inverter with DC disconnect

4. AC disconnect

5. PV production meter (if required by your electric utility)

6. Main electrical service panel (household breaker box)

7. Utility net meter (usually provided and installed by the utility)

Rooftop Array with Microinverters

This sample design is based on the actual PV system featured in the rooftop installation shown in chapter 6. It has 14 modules total, laid out in two separate rows on different roof planes. The upper row is on the main house roof and has 8 modules. The lower row is on a porch roof and has 6 modules. The transition between the two roof planes occurs directly above the rear wall of the house, as indicated on the drawing (the wall is below both roofs, and the roofing is continuous from the upper to lower planes).

The microinverters used here are designed to serve two modules each, so there is only one microinverter for every two modules, for a total of 7 microinverters. Each module connects to a microinverter, and the microinverters are designed to connect together — one to the next — rather than connecting to a separate trunk cable, as some microinverters do. A junction box is installed near the upper module support rails at the end of each of the two module rows. A wire lead from the first microinverter in the row goes into the junction box. The wiring runs from each junction box pass through conduit and down into the attic, where they are joined at a third junction box. From there, a single run of wiring extends down to the AC disconnect mounted on the house's exterior wall.

The microinverter specs shown give the values for an individual module (not a module pair). Comparing the module and microinverter specs:

Module V_{pmax} (**37.2** volts) must fit within the inverter's MPPT Voltage Range (**22 - 45** volts)

Module V_{oc} (**46.2** volts) must be lower than the inverter Maximum System Voltage (**55** volts)

Module I_{pmax} (**8.48** amps) must be lower than the inverter Maximum DC Input Current (**12** amps)

Note that the AC output of the inverters is **240-volt** (AC). This is typical for microinverters. The Maximum Units per Branch Circuit (**7**) means that you can connect up to 7 microinverters together and connect them to a 20-amp breaker in the main electrical service panel. In this design, there is only one branch circuit containing all seven microinverters.

DC–AC MICROINVERTER SPECS

Manuf.: APSystems   Model: YC500A

MPPT Voltage Range: 22-45 Volts

Max. System Voltage: 55 Volts

Max. DC Input Current: 12 Amps

AC Output Voltage: 240 Volts

AC Output Current: 2.08 Amps

Max. Units per Branch Circuit:
7 per 20 Amps @ 240 Volts

PV MODULE SPECS

Manuf: JinkoSolar   Model: JKM315P-V

P_{max} = 315 Watts

V_{pmax} = 37.2 Volts

V_{oc} = 46.2 Volts

I_{pmax} = 8.48 Amps

I_{sc} = 9.01 Amps

Length = 77.0 Inches

Width = 39.1 Inches

4.4 kW

Junction Box

36' 7"

South

12' 2"

Tilt: 21°

Rails

House Wall

Junction Box

= Microinverter

11' 8"

Tilt: 13°

Rails

AC Disconnect | PV Production Meter | Main Electrical Service Panel | Utility Net Meter

**PV Array Layout: 1 row of 8 modules (upper roof),
1 row of 6 modules (lower roof)**

SYSTEM ELECTRICAL SPECS

Module Orientation: Portrait Azimuth: 175° Tilts: Upper Roof 21°; Lower Roof 13°

DC Power System Rating: 14 Modules × 315 Watts = 4,410 Watts (DC, peak sunlight)

2 PV Modules per YC500A (DC–AC) Microinverter => 240 Volts @ 2.08 Amps (AC output each)

7 Total YC500A Microinverters in System
1 Branch Circuit: 20 Amp AC Breaker

TOTAL SYSTEM OUTPUT (PVWatts)

4,410 Watts (DC, peak sunlight)

6,248 kWh (AC, annual)

Ground-Mount, Landscape

This sample system demonstrates a ground-mount array with the modules in landscape orientation. At 15 kW, it's a relatively large residential system but shares many features with the basic rooftop system shown in Sample Design 1. Both systems use the same models of module and inverter, making most of the electrical calculations the same. The main difference is that this large system has 10 modules in each series-string (instead of 9) and three string-inverters instead of one so that each collection of modules has its own inverter.

By the way, with ground-mount systems, you can choose to mount the inverter(s) underneath the backside of each row or, alternatively, run DC wiring in a conduit in a trench back to the house. Clearly, there are pros and cons to each approach. If the inverter is mounted at the array, then you will be running AC wire back to the house. However, in this case, the inverter is outdoors and must be in an outdoor-rated enclosure (typically rated as NEMA 3R). If the inverter is located at the house (in the garage or basement, for instance) it will be out of the weather, but in this case you need to run DC wiring from the array to the inverter, which may be a considerable distance away and therefore may require heavy-gauge wire to minimize voltage losses over that distance.

DC-AC INVERTER SPECS

Manuf.: Fronius

Model: IG Plus 5.0

MPPT Voltage Range: 230–500 Volts

Max. Voltage: 600 Volts

Max. DC Input Current: 23.4 Amps

AC Output Voltage: 208/240/277 Volts

PV MODULE SPECS

Model: Helios 6T-250

P_{max} = 250 Watts

V_{pmax} = 30.3 Volts

V_{oc} = 37.4 Volts

I_{pmax} = 8.22 Amps

I_{sc} = 8.72 Amps

Length = 66.1 Inches

Width = 39.0 Inches

15.0 kW

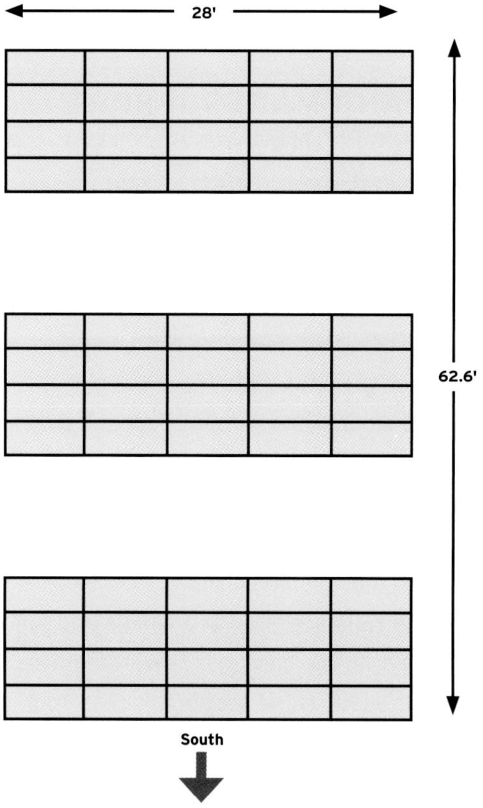

28'

62.6'

South

**PV Array Layout:
3 rows of 20 modules each**

MECHANICAL SPECS

Assume W–E post separation of 8 feet

20	# of Modules/Row	2	# of Series-Strings/Row
3	# of Rows	30	# of Posts
6	# of Base Rails	30	# of Module Rails
28	Base Rail Length (feet)	13.2	Module Rail Length (feet) (Hypotenuse)

Row Layout:

4	High	60	# of End Clamps
5	Across	90	# of Mid Clamps
		60	# of WEEBs
		13 Feet	Row-to-Row Spacing
		6.7 Feet	Peak Row Height above Ground
		5.2 Feet	Array Height

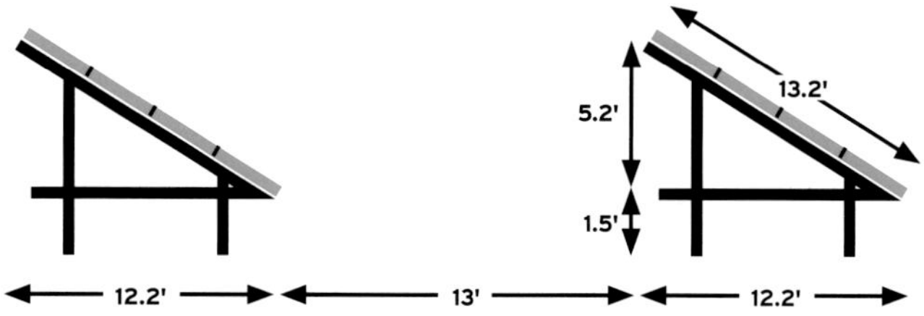

SIDE VIEW

SYSTEM ELECTRICAL SPECS

Module Orientation: Landscape Azimuth: 180° Tilt: 23°

60 x-Si PV Modules × 250 Watts = 15,000 Watts (DC, peak sunlight)

10 Modules per Series-String => 303.0 Volts @ 8.22 Amps = 2,491 Watts / String
(String Voc × 1.25 = 468 Volts)   (String Ampacity = 13.63 Amps)

2 Parallel Strings per **5kW** Inverter => 303.0 Volts @ 16.44 Amps = 4,981 Watts / Inverter
3 × **5kW** DC-to-AC Inverters => 14,944 Watts (DC, peak sunlight)   PVWatts: 21,771 kWh (AC/year)

TOTAL SYSTEM OUTPUT: 15,000 Watts (DC, peak sunlight); 21,771 kWh (AC, annual)

ROW-TO-ROW SPACING

Note that with this system, a "row" is not just a line of adjacent modules; it also refers to an entire grouping of modules on a single mounting structure. This system has three array rows, each four modules high by five across. You'll see this terminology used with specs for *row-to-row spacing* and *peak row height*. Proper row-to-row spacing ensures that multiple rows of modules don't shade one another. The calculation follows the 2½ times rule described in chapter 2 (page 33). In this case, the minimum spacing is determined by the height of the array (5.2 feet):

5.2 feet × 2.5 = **13** feet

Note that while the array height above the ground (called the *peak row height*) is 6.7 feet, the bottom edges of all the arrays are 1.5 feet above the ground to prevent shading by snow or vegetation. Only the height above this level contributes to row-to-row shading. Therefore, you use only the height of the array – not the array height above the ground – to calculate the separation needed to prevent row-to-row shading.

The 2½ times rule prevents shading when the sun is at its lowest in the sky (around December 21). If you're short on space, you can move the rows closer together and live with a little midwinter shading, or decrease the tilt angle to allow the rows to be closer together.

ARRAY AND MAIN SUPPORT CALCULATIONS

As with the design of a rooftop racking system, you'll work with the manufacturer of your ground-mount structure to determine the mechanical specifications for the mounting system and compile a parts list. With a landscape layout, the **base rails** (see Assembling the Ground-Mount Structure in chapter 7, page 132) run horizontally across the vertical support posts and extend the full length of the array row: in this case, five modules long, plus spacing between modules (typically ⅛ to ¼ inch). The **module rails** are tilted at an angle and run perpendicular to the base rails. Their length is equal to the width of the row: in this case, four modules high, plus spacing between modules (typically ¼ to ½ inch) for mid clamps and any additional space needed for end clamps.

ARRAY DIMENSIONS

The length of each array row (east–west dimension) is the total length of five modules, plus the spaces between adjacent modules:

5 (modules across) × **66.1** inches (module length) + module spacing = **28** feet

The width (or depth) of the entire array (north–south dimension) is the width of each array row times 3, plus two row-to-row spacings:

12.2 feet (row width) × **3** (array rows) = **36.6** feet

36.6 feet + **26** feet (**13**-foot row spacing × **2**) = **62.6** feet

NUMBER OF POSTS

Divide the base rail length by the post spacing (provided by professional engineer, usually after performing a soil test):

28 feet (base rail) ÷ **8** feet (post spacing) = **3.5** (round up to next whole number: **4**)

Add one for additional end post:

4 + 1 = **5** posts per post row/base rail

5 posts (rear row) + **5** posts (front row) = **10** posts total per array row

10 × **3** (array rows) = **30** posts total

NUMBER OF BASE RAILS

Count two base rails for each array row (one across the rear posts, one across the front posts):

2 × **3** (array rows) = **6** base rails total

NUMBER OF MODULE RAILS

Each array row has five vertical columns of modules. Each column needs two module rails:

2 (module rails) × **5** (columns) = **10** rails per array row

10 (per row) × **3** (array rows) = **30** module rails total

END CLAMPS

4 (end clamps) × **5** (modules wide) × **3** (array rows) = **60** total

MID CLAMPS

6 (mid clamps) × **5** (modules wide) × **3** (array rows) = **90** total

WEEBs

4 (2 WEEBs per 2 modules) × **5** (modules wide) × **3** (array rows) = **60** total

Ground-Mount, Portrait

This ground-mount system is identical to Sample Design 3, except for the module orientation. Note that the portrait layout here yields a row length of 33 feet (versus 28 with landscape) and a total array length of 52.2 feet (versus 62.6 feet with landscape). The shorter array length with portrait is due to shorter array height, requiring less spacing between rows (see Row-to-Row Spacing on page 78). The other differences lie in the rail orientation. With modules in portrait, the base rails run at an angle from atop the long rear posts to the top of the short front posts (see Assembling the Ground-Mount Structure in chapter 7, page 132). The module rails run horizontally along the full length of the row, secured to the tops of the base rails.

DC–AC INVERTER SPECS

Manuf.: Fronius

Model: IG Plus 5.0

MPPT Voltage Range: 230-500 Volts

Max. Voltage: 600 Volts

Max. DC Input Current: 23.4 Amps

AC Output Voltage: 208/240/277 Volts

PV MODULE SPECS

Model: Helios 6T-250

P_{max} = 250 Watts

V_{pmax} = 30.3 Volts

V_{oc} = 37.4 Volts

I_{pmax} = 8.22 Amps

I_{sc} = 8.72 Amps

Length = 66.1 Inches

Width = 39.0 Inches

15.0 kW

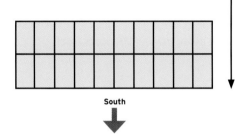

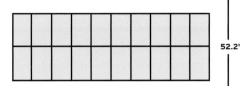

South ↓

PV Array Layout:
3 rows of 20 modules each

33'

52.2'

MECHANICAL SPECS

Assume W–E post separation of 8 feet

20	# of Modules/Row	2	# of Series-Strings/Row
3	# of Rows	36	# of Posts
18	# of Base Rails	12	# of Module Rails
11.1 Feet	Base Rail Length (Hypotenuse)	33 Feet	Module Rail Length

Row Layout:

2	High	24	# of End Clamps
10	Across	108	# of Mid Clamps
		60	# of WEEBs
		10.8 Feet	Row-to-Row Spacing
		5.8 Feet	Peak Row Height above Ground
		4.3 Feet	Array Height

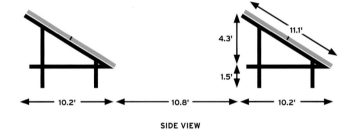

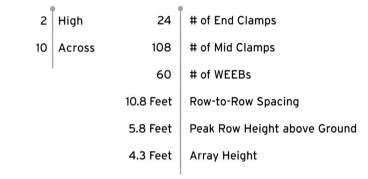

SIDE VIEW

SYSTEM ELECTRICAL SPECS

Module Orientation: Portrait Azimuth: 180° Tilt: 23°

60 x-Si PV Modules × 250 Watts = 15,000 Watts (DC, peak sunlight)

10 Modules per Series-String => 303.0 Volts @ 8.22 Amps = 2,491 Watts / String
(String Voc × 1.25 = 468 Volts) (String Ampacity = 13.63 Amps)

2 Parallel Strings per **5kW** Inverter => 303.0 Volts @ 16.44 Amps = 4,981 Watts / Inverter
3 × **5kW** DC-to-AC Inverter => 14,944 Watts (DC, peak sunlight) PVWatts: 21,771 kWh (AC/year)

TOTAL SYSTEM OUTPUT: 15,000 Watts (DC, peak sunlight); 21,771 kWh (AC, annual)

ARRAY AND MAIN SUPPORT CALCULATIONS

As with the other ground-mount design (Sample Design 3), this design with portrait orientation includes three rows of 20 modules each. The modules are arranged 2 high by 10 across in each array row. The electrical calculations remain the same.

ARRAY DIMENSIONS

The length of each array row (east–west dimension) is the total width of 10 modules plus the spaces between adjacent modules for mid clamps and space for end clamps:

10 (modules across) × **39** inches (module width) + module spacing = **33** feet

The depth of the entire array (north-south dimension) is the width of each array row times 3 plus two row-to-row spacings:

10.2 feet (row width) × **3** (array rows) = **30.6** feet

30.6 feet + **21.6** feet (**10.8**-foot row spacing × **2**) = **52.2** feet

NUMBER OF POSTS

Divide the module rail length by the post spacing:

33 feet (module rail) ÷ **8** feet (post spacing) = **4.13** (round up to next whole number: **5**)

Add one for additional end post:

5 + **1** = **6** posts per post row

6 posts (rear row) + **6** posts (front row) = **12** posts total per array row

12 × **3** (total array rows) = **36** posts total

NUMBER OF BASE RAILS

Count one base rail for each pair of front and rear posts:

6 (pairs per array row) × **1** (base rail) × **3** (array rows) = **18** base rails total

NUMBER OF MODULE RAILS

Count two module rails (each assembled from standard rail lengths) for the entire length of the array row (as with the rooftop array); each module row needs two rails:

2 (module rails) × **2** (modules high per array row) = **4** rails per array row

4 × **3** (array rows) = **12** module rails total

One-Line Electrical Diagrams

The **one-line diagram** is a simple drawing that shows all of the electrical elements in a PV system. It is a standard requirement for obtaining a permit for a PV installation. The elements are laid out in the correct location for the installation and are linked by a line indicating the wiring that will connect all the parts, plus an additional line or lines for the system ground wire. A set of notes providing specific details about materials, construction, design, codes, and signage helps keep the diagram clean and easy to read.

The local building department will have its own formatting and requirements for the one-line diagram and will likely have a sample that you can follow. The samples shown on the following pages are the one-line diagrams for Sample Design 1 (page 66) and Sample Design 2 (page 74). Typical elements required for one-line diagrams include the following:

- Name and address of customer and name of utility provider
- All electrical components of the PV system
- Key specs for PV modules and inverter(s)
- Conduit and wiring specifications between adjacent components

- Junction/combiner boxes
- Disconnects (DC and AC)
- Meters (production meter, utility net meter)
- Main electrical service panel
- Ground wiring and specifications

1-LINE ELECTRICAL WIRING DIAGRAM — 4.5 kW PV SYSTEM WITH STRING INVERTER

MODULE/STRING SPECS

18 HeliosSolar x-Si PV Modules

250 Watts (DC, peak sunlight)

2 Parallel String(s) of

9 Modules in Series per String

V_{pmax}= 30.3 Volts => 272.7 Volts/String

V_{oc}= 37.4 Volts => 420.8 Maximum String V_{oc}

I_{pmax}= 8.22 Amps => 16.44 Total Amps (DC)

I_{sc}= 8.72 Amps => 13.63 String Ampacity

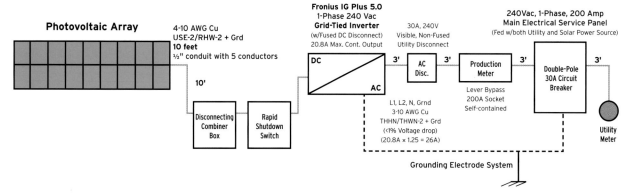

Grid-Tied, NO Batteries

Photovoltaic Array

4-10 AWG Cu
USE-2/RHW-2 + Grd
10 feet
½" conduit with 5 conductors

10'

Disconnecting Combiner Box

Rapid Shutdown Switch

Fronius IG Plus 5.0
1-Phase 240 Vac
Grid-Tied Inverter
(w/Fused DC Disconnect)
20.8A Max. Cont. Output

DC

AC

30A, 240V
Visible, Non-Fused
Utility Disconnect

AC Disc.

L1, L2, N, Grnd
3-10 AWG Cu
THHN/THWN-2 + Grd
(<1% Voltage drop)
(20.8A × 1.25 = 26A)

Production Meter

Lever Bypass
200A Socket
Self-contained

240Vac, 1-Phase, 200 Amp
Main Electrical Service Panel
(Fed w/both Utility and Solar Power Source)

Double-Pole
30A Circuit
Breaker

Utility Meter

Grounding Electrode System

Notes:

(1)   Minimum Permanent Signage (Yellow Placards, Black Lettering):
    a. On front cover of Main Service Panel: "WARNING: This panel also fed by Solar Electric Source"
    b. On the meter socket or immediately adjacent: "PHOTOVOLTAIC SYSTEM CONNECTED"
    c. On visible utility disconnect: "PV AC DISCONNECT"

(2)   PV Production Meter:
    a. Single-phase self-contained meter sockets for the utility PV production meter is required to be lever bypass
    b. Self-contained PV production meter is required to have the PV generation wired to the line-side terminals (top of meter block) so that standard meters can be utilized. The production meter will be installed in compliance with the March 1, 2011 revision of the Xcel Energy Standard, specifically Section 4, pages 44–47.
    c. Signage: "PV PROD" and customer address (stamped brass, aluminum, or stainless steel)

(3)   Signage other than that given in (1) and (2) above will be NEC 705.10 compliant

(4)   All components – PV modules, string combiners, GFI, DC disconnect, DC–AC inverter, AC disconnect, and utiltity disconnect – are UL-Listed

(5)   DC surge protection in each DC disconnect; AC surge protection included

(6)   Lightning and equipment grounding systems are bonded together

(7)   Interconnection at double-pole 30-amp circuit breaker in main service panel

(8)   System shall not be tested or put on line until approved by utility

(9)   Photovoltaic System will be installed in compliance with Article 690 of the NEC

System:	Helios Solar Works / Fronius Photovoltaic System
Name:	ABC
Designer:	B.T.U., LLC, Lakewood, CO
Installer:	B.T.U., LLC, Lakewood, CO
Date:	2016

MODULE SPECS

14 Jinko Solar x-Si PV Modules: 315 Watts (DC, peak sun)

7 DC-AC Microinverters (1 per 2 modules)

Modules:

V_{pmax} = 37.2 Volts

V_{oc} = 46.2 Volts

I_{pmax} = 8.48 Amps   Conductor Ampacity:

I_{sc} = 9.01 Amps   14.08 Amps

MICROINVERTER SPECS

Model: APS YC500A

MPPT: 22-45 Volts

Max. Voltage: 55 Volts

Max. DC Input: 12 Amps

AC Output: 240 Volts (AC)

2.08 Amps (AC)

Grid-Tied, NO Batteries

240Vac, 1-Phase, 200 Amp
Main Electrical Service Panel
(w/Fused DC Disconnect)
20.8A Max. Cont. Output

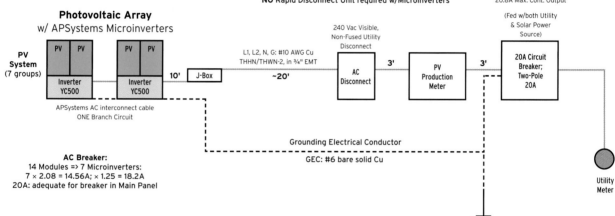

PV array has a single AC branch circuit
so NO Solar Subpanel/AC Combiner is required;
NO Rapid Disconnect Unit required w/Microinverters

AC Breaker:
14 Modules => 7 Microinverters:
7 × 2.08 = 14.56A; × 1.25 = 18.2A
20A: adequate for breaker in Main Panel

Notes:

(1) Minimum Permanent Signage (Yellow Placards, Black Lettering):
 a. On front cover of Main Service Panel: "WARNING: This panel also fed by Solar Electric Source"
 b. On the meter socket or immediately adjacent: "PHOTOVOLTAIC SYSTEM CONNECTED"
 c. On visible utility disconnect: "PV AC DISCONNECT"
 d. All Signage (stamped brass, aluminum, stainless steel, engraved plastic) will be NEC 705.10 compliant

(2) PV Production Meter:
 a. The production meter is required, per Xcel Energy regulations

(3) Signage other than that given in (1) and (2) above will be NEC 705.10 compliant

(4) All components – PV modules, string combiners, GFI, DC disconnect, DC–AC microinverters, AC disconnects, and utiltity disconnect – are UL-Listed

(5) DC surge protection in each DC disconnect; AC surge protection included

(6) Lightning and equipment grounding systems are bonded together

(7) Each branch circuit interconnection is a double-pole 20A or 25A circuit breaker

(8) System shall not be tested or put on line until approved by utility

(9) Photovoltaic System will be installed in compliance with Article 690 of the NEC

System:	Jinko Solar / APSystems Photovoltaic System
Name:	ABC
Designer:	B.T.U., LLC, Lakewood, CO
Installer:	B.T.U., LLC, Lakewood, CO
Date:	2016

Wire Types for PV Systems

Most residential PV systems use three different types of wire: one type for the array, one type for running inside the conduit, and one type for the ground wires. Note that the **type** of wire or cable is not the same thing as the **size** of the conductors. The conductor is the metal portion of the wire, exclusive of any insulation or sheathing. Electrical wire may contain one or more solid conductors (a single metal wire) or stranded conductors, made of smaller wires wound together.

WIRE TYPES AND CHARACTERISTICS

	WIRE TYPE	CHARACTERISTICS
	PV wire	Single-conductor, insulated wire for outdoor exposure. Rated for 90°C (184°F) in wet locations and up to 150°C (302°F) in dry locations. Stranded copper conductor. UV-protected sheathing. Use red for DC positive (+), black for DC negative (−).
	USE-2	Single-conductor, insulated wire for outdoor exposure. Rated for 90°C (184°F) for wet and dry locations. Solid or stranded copper, copper-clad aluminum, or aluminum conductor. UV-protected sheathing.
	THWN-2	Insulated wire rated for 90°C (184°F) for wet and dry locations. Solid or stranded. Not UV-protected.
	Ground wire	Copper ground wire. Bare (uninsulated) solid conductor or insulated green stranded conductor.
	UF cable	Underground feeder cable for direct burial. Solid-copper conductors.

TYPE generally relates to the wire or cable construction, indicating materials (such as insulation) and suitability for specific uses.

SIZE is indicated by the gauge, or diameter, of the conductor (not including insulation, cable sheathing, or other materials). In North America, the standard wire gauge system is American Wire Gauge (AWG). Wire gauge indicates the current-carrying capacity of a conductor. The smaller the gauge number, the larger the wire's diameter.

The chart below shows the main types of wire/cable and size used for most residential PV systems. Work with your electrician to determine exactly what type and size of wire to use for your system (and your one-line diagram).

You may have to change wire types and/or sizes as you make all the connections from the PV array to the main electrical service panel. A good inspector likely will check to see if your actual installation agrees with your one-line diagram before passing the system on the final inspection.

TYPICAL SIZE FOR RESIDENTIAL PV	WHERE TO USE
10 AWG	**Home run wires** and other exposed connections for PV array. May be preferred over USE-2 for exposed runs. Does not need to be installed in conduit.
10 AWG	Alternative to PV wire but less heat-tolerant. Solid-conductor types not as flexible as stranded-wire conductor. Does not need to be installed in conduit.
10 AWG	Connections from array to ground-level components, and between ground-level components. Must be installed inside conduit.
6 AWG	Bare solid wire used for grounding metal parts of array and for exposed runs to system ground. Stranded insulated wire used inside conduit.
10 AWG	Underground runs between ground-mount arrays and house. May be run inside conduit, but conduit is typically not required for the portions of cable at the specified burial depth.

5 Getting Ready to Install

GOAL

Obtain a permit, buy equipment,
and prepare for the installation

WITH YOUR DESIGN DOWN ON PAPER, you're ready to move from the drawing board to the building department, where you'll declare your intention to become an energy producer! (Actually, you can do this online in many places, which may lack drama but saves on gas, not to mention having to get out of your pajamas.) The most important step in this phase is getting your permit. Without it, your DIY project is DIW (dead in the water). With the city's (and utility's) approval in hand, you can make your equipment purchases and begin applying for rebates and other financial incentives. The final (prep) step is assembling your crew and preparing for installation day (or installation weekend, at least for the array). The more helpers the better, but it doesn't have to look like a barn raising; two or three strong assistants can get you through the most physical part — lifting and placing the modules.

Permitting and Inspections

Obtaining a permit for a PV installation can take as little as a few hours to upward of a few weeks. The utility may take even longer to approve the system for rebates. It all depends on your project and, more important, the local government's policies regarding solar installations. While many authorities have adopted expedited permit processes, others are less accommodating to renewables. If you're dealing with the latter, try to be patient, and take credit for helping pave the way for future homeowners who, hopefully, will face less resistance someday when going solar. As for the utilities, they're required by law to buy your solar-generated electricity, but they're not required to make it easy.

The Permit Packet

A permit packet is a document or set of documents and forms that outlines the required information for obtaining a building permit for a PV installation. In some cases, a packet may include a permit application and all necessary supporting documents to apply for a permit. In other cases, a packet may serve only as a checklist to help you prepare an application, and several additional documents or forms may be required.

Here are some of the common items included or listed in a PV permit packet:

◇ **Site plan:** Aerial view of your entire property, including access roads, driveways, property lines, and orientation (north, south, etc.).

◇ **Array layout:** A simple drawing of your rooftop or ground-mount area and relevant surrounding property, showing the solar array and module layout, locations of disconnects, inverters, meters, and home electrical service panel. May include roof penetrations for array wiring.

◇ **Rooftop load calculations:** Worksheet for totaling the weight of modules and racking, broken down to a load rating of pounds per square foot (psf). May include a description of roof framing (rafter/truss with spacing dimension) and roofing (roofs with more than one layer of roofing may require additional inspection and approval). The worksheet may need an engineer's stamp confirming load calculations. A letter from the architect or builder also may be required for PV systems that are part of new construction projects.

◇ **Ground-mount design:** Construction details for the ground-mount structure, such as elevation drawings, concrete footing/pole anchor details, and mounting structure specs. Stamped engineered drawings are often required but may be supplied by the ground-mount manufacturer.

◇ **Elevations:** Simple elevation drawings, sometimes required for ground-mount structures and for rooftop systems that are not flush-mount, including flat-roof systems.

◇ **Zoning permit/approval:** Most commonly needed for ground-mount structures and for nonstandard or non-flush-mount rooftop systems.

◇ **Engineer's drawings:** Stamped plans often required for wind, snow, and load ratings on both rooftop and ground-mount arrays.

◇ **One-line electrical diagram:** See page 83 for common requirements.

◇ **Hardware spec sheets:** For modules, inverters, module mounting structure, electrical components, and so on.

- **Electrical calculations:** String and system voltage, amperage, and wattage calculations for either string inverter(s) or microinverters, as applicable.

- **Utility approval:** Confirmation from utility that it will hook up your system. Often required before applying for permit.

- **Description of** warning signage and markings required for final inspection, noting disconnects, rapid-shutdown devices, presence of multiple power sources, and so on. (May be indicated on the one-line diagram.)

- **Master electrician's credentials:** Including your electrician's name, company, license number, and contact info. May also require proof of NABCEP certification for solar professionals on the project.

Because permit requirements and processes vary from city to city or county to county, and the only one that matters is your own, it's best to dive right in and download a copy of the packet or checklist (or similar document) from the local building department. This is also a good time to learn about permitting costs, so you'll know what to expect. Costs, like permitting requirements, vary widely – from under $100 to $300 or more. Many authorities assess fees as a percentage of the project cost, as is done with other home improvement and remodeling permits.

Once your application is approved and you've paid the fee and obtained a permit, you can legally begin installing your PV system. Of course, if you're following the recommendations in this book, you have yet to buy your equipment. You can paint yourself into a corner if you buy your hardware before everything is approved.

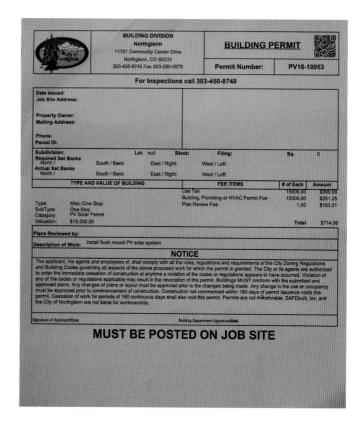

Inspections

The building department will tell you how many inspections you need for your project and how and when to schedule them. Pay close attention to the specific requirements for each inspection to be sure you have everything ready when the inspector arrives, and have the entire permit packet handy in case he wants to see portions of it. In addition to showing the inspector the work you've done, you'll likely need the following:

- Your project permit (enclosed in plastic and visibly displayed) and utility approval

- Manufacturer's plans, specs, and installation instructions for PV hardware

- Warning/safety signage and labels

- An OSHA-compliant ladder secured to your roof (for rooftop systems)

Standard PV installations commonly entail two inspections — a rough and a final — and sometimes an intermediate inspection as well. If all parts of the system are visible and accessible, such as with most ground-mount systems, you may be able to combine inspections into one final one.

NOTE: *It's possible that you'll need or want to make changes to your system design and/or installation plan after you get your permit. Changes are commonly allowed, but you should always get the changes approved by the building department — with a new stamp, if applicable — before proceeding with the new plan. Have the new documents on hand for the final inspection. You don't want to have to tell the inspector that you made changes on your own and failed to get approval.*

TIP

SHOW YOUR WORK

Like math teachers, inspectors want to see your work. Therefore, before you start your installation, it's critical that you know exactly when to call for inspections. With rough inspections for rooftop systems, for example, inspectors may want to visit when only some of the modules are installed, so they can see the racking, how it's flashed and attached to the roof, and so on. If you cover up what the inspector needs to see, you may have to undo some of your work to pass an inspection. For ground-mount systems, you will want to take a photo of a tape measure in the hole (and trench) to show the inspector (see Photo Evidence, page 130).

Shopping Solar

For an industry with a limited history of DIY installation, the solar equipment market is surprisingly DIY-friendly. Online retailers have made it easy to compare products and check manufacturer specs, and pricing is equally straightforward. Modules and PV system kits are sold widely, even through places like hardware stores, home centers, and online mega-retailers. But it's best to stick with solar specialists, companies that are fully invested in solar and are set up to supply equipment for a range of projects, from residential to larger commercial systems. These retailers offer a large selection, competitive pricing, and knowledgeable support staff who can answer your technical questions. Solar specialists' industry knowledge also will come in handy if you ever need to make a warranty claim with a manufacturer.

That's not to suggest that you buy exclusively from online retailers. Before making a big purchase, you should also look at local manufacturers, local suppliers and distributors, and even local solar installers. Here's a suggested process:

1. Shop online to get a sense of the product selection and pricing in the general market.

2. Look for nearby manufacturers of major hardware: modules, racking, inverters. If any are close enough, you could pick up the goods yourself and save on shipping costs. Plus, you'd be buying local and supporting the domestic solar industry.

3. Ask about pricing and shipping options through local suppliers and distributors. This is a way to keep some of your dollars local and may reduce shipping costs.

4. Talk with local professional solar installers about buying equipment through them. They might pass part of their pro discount on to you while still profiting from the sale. In addition, they might offer recommendations and advice for your product selection as well as system design and installation tips.

5. Show your final design plans to the supplier to make sure your hardware delivery is accurate (model, size, quantity).

SHIPPING ADVICE

A pallet of expensive PV components is not something you want to have sitting out on the curb, lest it get swiped or damaged before you get home from work. Make sure you're around to receive all shipped items and to inspect the containers before you sign for the delivery.

GET READY FOR LARGE SHIPMENTS. Prepare a safe and secure (lockable) storage space for your PV components. The delivery driver can unload your pallet but probably can't bring it to your garage, so be prepared to break down the shipment and carry everything in right away. Ask the shipper whether the driver will have a lift gate and/or pallet jack to simplify unloading the pallet from the truck and, if possible, moving it to a convenient and secure location.

INSPECT ALL PACKAGES FOR DAMAGE. If a box has a crushed corner, tear, hole, or any other damage, photograph the box from all directions; this will help immensely when you make a claim for damaged goods.

CHECK FOR MISSING ITEMS. Compare the manufacturer's shipping label to the box labels and the actual contents of the shipment.

REPORT ALL DAMAGED OR MISSING ITEMS ASAP. Retailers give you only a day or two (at best) to report problems with shipments.

Local Materials

Chances are, you'll have to order only the major PV components: modules, support structures, inverters. The rest of the materials should be available through local suppliers and retailers. Things like screws, lag bolts, sealant, wire ties, and odds and ends are easy to find at any hardware or big-box store. For bigger (and pricier) items, like wiring and support poles for ground-mount structures, call commercial supply houses for pricing. Commercial suppliers usually sell to electricians and other pros, but many will also sell to nonprofessionals who know what they need. You might have your electrician supply all the electrical hardware (such as boxes, disconnects, breakers, etc.) and this may be cheaper than buying it yourself.

Your local shopping list may include the following:

◇ Fasteners for footers, racking, and electrical components

◇ Roofing sealant (use high-quality roofing sealant, not silicone or other types of caulk)

◇ Materials for ground-mounting (poles, gravel, concrete, and related items)

◇ Wiring and wiring connectors

◇ Conduit, along with clamps, fittings, and flashing/boots (for roof penetration)

◇ Electrical boxes, breakers, disconnects, meter boxes (ask your electrician)

◇ Mesh or netting for critter-proofing the array

◇ Pizza and beer (for the helper crew; quality counts here, too)

Solar Financials

As home improvements go, a PV system might be considered a midrange project in terms of initial cost. It's similar to replacing an asphalt-shingle roof or replacing several key appliances. These have value in the service or protection they provide, but only a PV system offers a monetary return on your investment by lowering your electric utility bills for the life of the system (and only roofing can rival a PV system for warranty life and low maintenance).

To provide a **return-on-investment (ROI)**, the thing you've invested in either has to increase in value or produce something that has value. PV systems do the latter, of course, but a true ROI isn't realized until the system has paid for itself; this is what is known as **payback**.

Pretty Simple Payback Calculation

There are many ways to describe payback on a PV system, but perhaps the simplest is the idea that you're paying for a certain number of years of electricity up front. If you divide the PV system cost (minus rebates and credits) by your average annual electricity savings, you'll get the number of years it will take for the system to pay for itself. After that period is up, your electricity is largely free (since there are almost no maintenence costs) and therefore represents a positive ROI. This calculation works fine if you consider a few additional factors. Here's a simple example:

◇ You pay $0.10 per kWh for grid power and use an average of 8,000 kWh per year. Your annual electricity bill is $800. (Do not count regular utility service charges, which will continue after the payback period).

◇ You install a 5 kW PV system to cover 100% of your average annual usage. The PV system costs $7,000 after rebates and credits (assuming DIY installation). The "simple payback" calculation is:

$$\$7,000 \div \$800 = 8.75 \text{ years}$$

However, here's why that's not the full picture for most grid-tied systems: Utilities typically don't buy your solar-generated power at the same rate that they sell you power from the grid. Back to our example: You're paying $0.10 per kWh for grid power, but the utility pays you only $0.03 per kWh for the excess power produced by your PV system. During daylight hours, when you produce more power than you use, you get a monetary credit of $0.03 for each kWh you produce over your usage. However, at night, when you use more power than you produce, you pay $0.10 per kWh.

This will add to your annual electricity cost after your system is in place, increasing the length of the payback period and continuing indefinitely after the payback period (and it will rise with utility rates). You can estimate the additional expense by looking at your current usage of electricity for each month of the year and comparing that to the expected solar production of your PV system for each month.

For example, in each month of June, July, and August (with longer days and higher solar irradiance) you use 400 kWh and your system produces 550 kWh, so you get $0.03 × 150 = $4.5 per month as a credit from the utility. By the end of the summer you're up $13.50. Let's assume September through November are flat; your usage equals your solar production. In December, January, and February (with shorter days and lower solar irradiance), you use 450 kWh each month and produce only 300 kWh, so you pay $0.10 × 150 = $15 for the extra grid power − a total of $45.00 for the season. If the remaining months are flat, you have an annual additional cost of $45.00 - $13.50 = $31.50.

To factor this into your payback (and ROI) calculation, you actually **subtract** the $31.50 from your annual usage cost above (because payback is based on what the PV system contributes, while the original $800 is based on your annual utility costs). Therefore, your payback period is calculated as follows:

$$\$7,000 \div \$768.50 \ (\$800 - \$31.50) = 9.11 \text{ years}$$

This example doesn't show a huge difference, but it does demonstrate how you can end up with a net expense for electricity (not counting those flat monthly fees for grid service), even if your PV system covers 100% of your annual power usage.

Financial Incentives

The strategy for getting financial incentives to help offset the cost of your PV system is simple: find out what's available in your area and go for everything you can get. You can find out what's available at the Database of State Incentives for Renewables & Efficiency (DSIRE), an up-to-date listing of federal, state, city/local, and utility-sponsored incentive programs and policies for renewable energy producers (soon to be you). Just click on your state or filter your search with your zip code to find programs that might apply to your project. Program listings include overview information and highlights of the terms as well as a link to the official sponsor site.

Programs, benefits, and eligibility vary by jurisdiction and by project (for example, system size can affect benefits), so it's important to learn all the details from the program sponsor. Three of the biggest and most widely available programs are local utility rebates, the federal tax credit, and Solar Renewable Energy Certificates (SRECs), all of which can have a significant financial impact on your PV system.

REBATES

Rebates can take many forms, but a common type is a one-time payment, based on your system size, paid to you by the electric utility. For example, if you install a 5 kW system and the rebate amount is $0.50 per watt, the total payment is $2,500. Large systems may be subject to a cap. The rebate is paid after the utility receives a copy of the approved final inspection for your system.

FEDERAL TAX CREDIT

The Residential Renewable Energy Tax Credit was extended by Congress in 2015 and runs through 2021. It's a federal tax break worth up to 30% of the total cost of a new PV system.

Simply put, if your total system cost is $10,000, you can get a one-time credit on your tax return of up to $3,000. If your tax liability for the first year the credit is taken is less than your eligible credit amount, you can carry over the remaining credit to the following tax year. The program is set to expire at the end of 2021. The maximum credit tapers to 26% for systems installed after 2019 and to 22% for systems installed after 2020. You apply for the credit with IRS Form 5695. Consult your tax adviser for program details and requirements.

SRECS

Solar Renewable Energy Certificates (SRECs) are certificates you get for producing solar-generated electricity. One SREC is awarded for every 1,000 kWh (1 megawatt-hour) produced. If your PV system produces 6,000 kWh a year, you get six SRECs that year, or an average of one every two months.

SRECs are awarded for solar-electricity production and have nothing to do with your household usage. You can use the power in your house, you can sell it back to the grid — it doesn't matter what you do with it; you still get the SRECs for producing the power.

SRECs are part of a state-level system that tracks solar electricity production to meet renewable energy goals. The system, known as the Renewable Portfolio Standard (RPS), varies from state to state. Many states have laws requiring that a portion of the electricity produced in the state comes from renewable sources. To meet this requirement, utilities can produce their own renewable energy, or they can buy renewable energy certificates (RECs). SRECs are just for solar; there are also RECs for wind and hydroelectric power.

What this means is that your SRECs have value. How much value depends on where you live and the condition of the SREC market, which is subject to supply and demand. Only a handful of states have active markets in which SRECs are publicly traded commodities. Homeowners with PV systems in these states typically sell their SRECs through a broker or aggregator, a trader or firm that buys and sells RECs much like a stock trader. Check with your state government or state renewable energy program to learn about SRECs in your area.

If your state doesn't have an active market, your SRECs still have value. You may be able to sell SRECs in markets that allow out-of-state credits, but most likely you'll sell them to your utility. One way is to contract with the utility for a fixed period (perhaps 20 years) so that you get a set amount for your solar production, paid by check each month (or you may be allowed to roll them over and get paid at the end of the year). For example, if you produce 500 kWh of electricity in a month and your contracted rate is $0.02 per kWh, the utility sends you a check for $10 (500 × $0.02) for that month. In this case, you don't have to produce a full megawatt before getting paid for the power. The utility's program might not even use the term *SREC*, but it's the same system.

The most important thing to know about SRECs is that **you must register your PV system** to have your energy production tracked and recorded so that you can receive the credits. It's best to register before you install your system, as restrictions may apply. For some utility customers, applying for SRECs is part of the *Utility Interconnection Agreement* that you complete as part of getting your system approved by the utility. Contact your utility and state government or energy agency for details. Further, the utility will not install a net meter or give you permission to operate until the Interconnection Agreement is signed.

Assembling Your Crew

While it's possible to install portions of a PV system by yourself, you'll need some helpers for at least a couple of days. For a rooftop installation, the hardest part is getting the modules up onto the roof and onto the racking. This works best with four people — two on the ground and two on the roof — but it can be done with two on the roof and one strong worker on the ground. Modules weigh about 40 to 60 pounds each and typically measure 39 inches wide by 65 or 77 inches long. It's best to lift them up to the roof and mount them one at a time. That means you'll need the full crew for the entire module installation; plan on one to three days, depending on the system size.

Prior to installing the modules, you'll need at least one helper for laying out and installing the racking; plan on a full day for that, too. Once the modules are up, it's best to have at least one helper for the conduit and wiring runs to the ground, but this can be a one-person job if necessary.

Time and labor for ground-mount arrays are subject to several variables, including the array size, the type of support system, and the installation site. Typically, the hardest part of the job is digging the holes for the concrete footings for the posts and the trench for the wiring to the house. You can speed up the work by renting a power auger for the footings and a trencher for the trench.

A well-fed crew is a happy crew.

If you will have a lot of footings and/or the digging will be difficult (your soil has a lot of clay or rocks), the footing prep might take two or more days. The more help you can get here the better, since it's just grunt labor, but one person can do the digging if necessary. After the holes are dug, you should have at least one helper to set the concrete posts and poles and pour the footings; plan on a day for that, plus three to four days of waiting time for the concrete to cure. In the meantime, you can work on other tasks, like digging a trench for underground wiring.

Once the concrete has cured, plan to have at least one helper for two to three days to construct the module racking and install the modules. After that, it's possible to complete the installation by yourself, but having a helper or two will speed the work.

Rooftop Safety

Everyone working on your roof should wear fall-arrest gear. This is the only safe way to protect against a fall, and it's not the place to cut

corners. So just bite the bullet and buy a setup for everyone in your crew who will be on the roof.

Basic fall-arrest systems start at around $100 and include a harness; a lanyard that connects the harness to the safety rope; the safety rope; a rope grab, or clutch, which allows for movement up and down the rope; and a roof anchor. Anchors are installed with nails or screws driven into roof rafters. Depending on the type, they can be located at the roof ridge or over any rafter, provided it's a minimum distance from the roof edge (typically 6 feet). Some installers like to place anchors on the opposite side of the ridge from the installation area to keep them out of the way. Anchors may be designed for temporary or permanent installation, and both types are reusable. Permanent anchors might come in handy for future module maintenance and cleaning. Be sure to seal any fasteners on permanent anchors, or any holes left by fasteners in temporary anchors, to prevent roof leaks.

Anchor screws

Anchor

Parts and setup of a basic fall-arrest system. You need a complete system for each worker, and only one person can use one anchor at a time.

Lanyard

Safety rope

Harness

99

Ladder Safety

A proper ladder setup is just as important as rooftop safety gear. Arrange to have one or two heavy-duty extension ladders for all rooftop work. You'll also need a ladder for the inspector to use for the rough inspection (and you can bet the inspector won't like to see a rickety old wooden job). You can always rent a good ladder or two if you don't want to buy one. Follow these essentials when setting up any ladder:

◇ Ensure the ladder feet are level. Compensate for uneven ground using the ladder's adjustable feet, or build up the low ground with plywood or a concrete block.

◇ Lean the ladder at a 75-degree angle, or 1 foot away from the wall for every 4 feet of height.

◇ Extend the ladder at least 3 feet above the roof.

◇ Tie off the ladder where it meets the roof to prevent it from sliding sideways. You can use wire, strong rope, or webbing secured to gutter anchors or tied to nails, screws, or eyebolts driven into the fascia board along the eave.

◇ Use extension ladders only. Never use stepladders for roof access or for lifting materials up to the roof; stepladders cannot be secured and can easily tip sideways.

◇ For ground-mount arrays, secure your ladder to the top-most horizontal rail.

3'

75°

Ladder feet
level and
secure

How to Get Modules onto the Roof

Hauling modules onto single-story roofs usually is no big deal, as the roof edge might be only 9 or 10 feet above the ground. In this case, most installers simply lift the modules from the ground, one at a time, and hand them up to two workers on the roof. Again, it helps to have two people on the ground as well, because modules aren't light (and they aren't cheap).

If a single-story roof is a bit too high for lifting from the ground, you can slide the modules up a ladder for a few feet until the rooftop crew can safely reach them. Just be aware of the hazards and strain involved, and don't climb the ladder higher than a few rungs. Alternatively, you can rent a level or two of scaffolding and pass the modules up in stages. To prevent marring the modules, always slide them faceup on the ladder, with the module frame resting on the ladder itself.

Multistory roofs require more planning. The best tool for the job is a boom lift (also known as a bucket lift, basket lift, or knuckle lift), which you can rent by the day for about $200 to $500 or weekly for about $1,000. This type of lift has an articulating boom that can extend and rotate to bring a load of modules not only up to the roof but also over and above the roof surface, so that workers don't have to unload from the roof's edge. The cheapest type is likely to be a towable, fixed-location model that stands on support legs. This usually works fine, because you don't need the mobility of a four-wheel-drive tractor-type boom. A boom lift needs a relatively flat surface for stability and at least one person who's learned how to operate it.

The other option for high roofs is rented scaffolding. A level of standard scaffolding is only about 5 feet tall, so you may need four or more levels for a two-story house. With this option there's no machinery to operate, but the rental may not be significantly cheaper than a boom lift, and you'll spend more time setting up and tearing down the scaffolding. This is also the manual approach, requiring more muscle power. Be sure to follow all recommendations when erecting and securing the scaffolding so that it is plumb and stable and protected and secured against tipping over.

A boom lift is a time- and back-saver that might not cost a lot more than scaffolding.

6 Mechanical Installation:
Rooftop

GOAL

Install the rooftop array, including racking, modules, microinverters or DC optimizers (as applicable), and grounding wires

RALLY THE CREW AND CINCH UP THOSE HARNESSES — it's solar time! This chapter walks you through the installation of a standard flush-mount PV array on a sloped roof with asphalt (composition) shingles. It applies to both grid-tied and off-grid systems. (In addition, there's an overview of flat-roof installation on page 124. Ground-mount installation is covered in chapter 7, but you'll come back here for the module details.) Of course, as you surely learned during the system design, PV hardware isn't exactly standard. While PV modules are relatively interchangeable, racking systems are highly product-specific. Every manufacturer has its own special doohickies and techniques for mounting and leveling the racking, securing the modules, grounding, and performing other essential functions. That's why your true guide for the installation will be the manufacturers' instructions and the local building code. If you have questions about your hardware or how to use it with your roofing material, contact the manufacturer or a solar installation professional.

Laying Out the Racking

This phase of the array installation consists of three stages:

1. Completing the basic layout of footers and rails: mapping the location of the racking and modules on paper

2. Locating rafters (or trusses) for footers: measuring for the precise locations of the roof framing members that you will anchor into

3. Snapping chalk lines and marking pilot holes: creating reference lines on the rooftop and marking pilot holes for installing the footers

The footers of your racking system are anchored into the roof rafters or trusses with lag screws. The size, length, and number of lag screws, as well as the number of footers, may need to be specified by an engineer, to satisfy the building department. These specs are based on the size and spacing of the rafters/trusses as well as the local building code requirements for wind and snow loads.

To ensure a strong connection, the lags must go into the meat of the framing, at the centers of the rafters. For beginners, the most accurate and foolproof way to locate rafters is to measure the rafter spacing from inside the attic (or from inside the house, if you have vaulted ceilings). Pros usually locate rafters from above the roof, using the old carpenter's trick of hammering and listening for the right sound. If you know how to do this effectively, you can use this technique instead. Just be careful to hammer only in areas that will be covered with flashing, as the hammer blows are hard on the shingles and you'll want the extra protection of the flashing to prevent further damage over time.

Before you climb onto the roof, make sure your ladder is tied off and everyone working on the roof is outfitted with fall-arresting equipment (see page 98). Never work on a roof that is slippery due to rain, snow, ice, or debris. Make sure your roof is dry and clean (no tree branches or leaves) before starting the installation.

Tools

- **Tape measure**
- **Drill**
- **¼-inch drill bit, 12 inches long**

Materials

- **¼- or ½-inch graph paper**
- **Pencil**
- **Ruler**
- **Straight wire or coat hanger**
- **Tape**
- **Marking crayon**
- **Stud finder (for finished/vaulted ceilings)**
- **Chalk line**

LAYING OUT THE FOOTERS AND RAILS

1. Create a rooftop map. Draw a new map on ¼- or ½-inch graph paper, or use a copy of the array layout from your system design. Include the roof dimensions from ridge to eave and from side to side (typically west to east). Also mark all of the rafters; their spacing doesn't have to be precise, but it must show the correct number of rafters in the installation area. Use one of the basic measuring techniques shown in Locating the Rafters (page 106) to find the rafter spacing (see Framing Dimensions, opposite); then use this spacing to lay out the rafters on your map. You will measure for the exact locations of the rafters in the next stage of the layout.

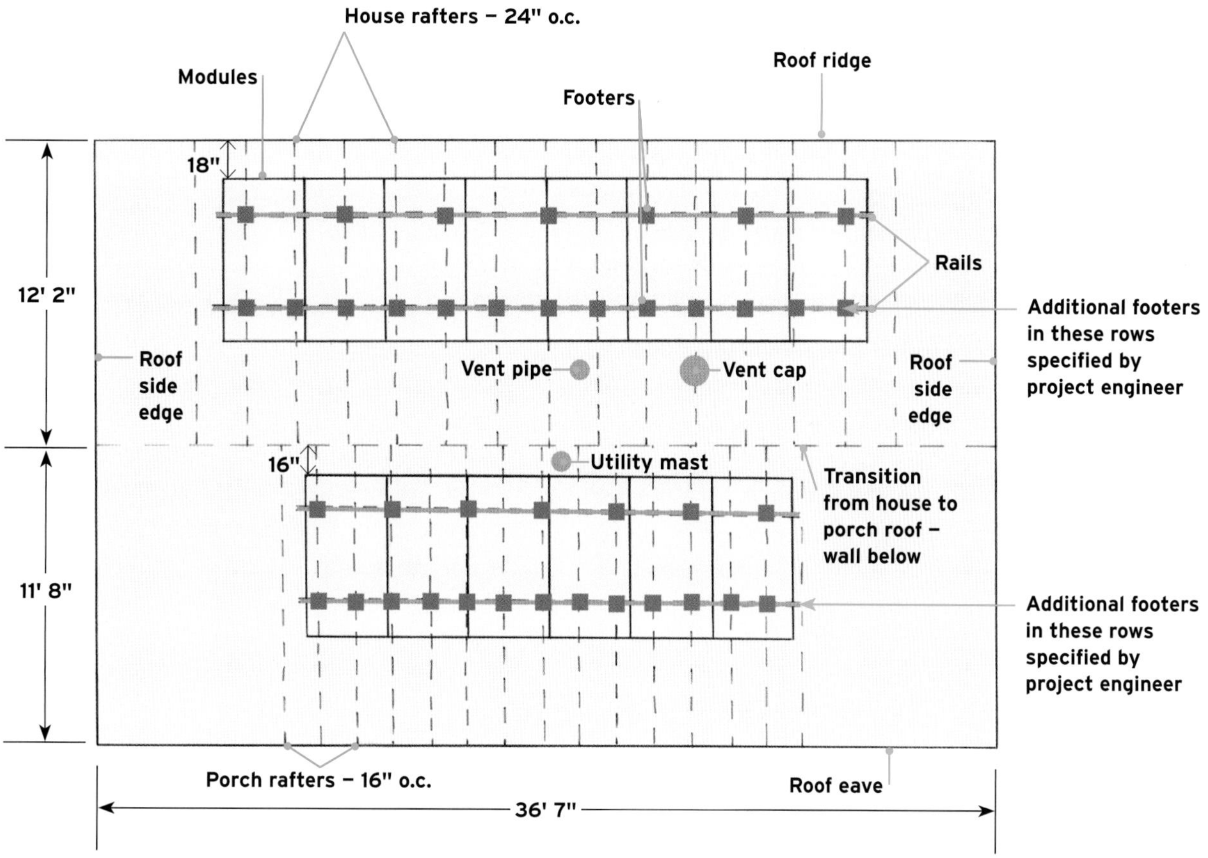

House rafters – 24" o.c.

Modules

Footers

Roof ridge

18"

12' 2"

Rails

Roof side edge

Vent pipe

Vent cap

Roof side edge

Additional footers in these rows specified by project engineer

16"

Utility mast

Transition from house to porch roof – wall below

11' 8"

Additional footers in these rows specified by project engineer

Porch rafters – 16" o.c.

Roof eave

36' 7"

TIP
FRAMING DIMENSIONS

Spacing of rafters, trusses, and other framing members typically follows *on-center* layout (abbreviated "o.c."), meaning they are spaced according to the distance between the center of one rafter and the center of the next (not the open space between the rafters). Most rafters are spaced 16 or 24 inches on center, while trusses are more commonly spaced 24 inches on center. Both rafters and trusses are typically made with "2-by" lumber (2×4, 2×6, etc.), which is about 1½ inches wide. When you measure or mark a layout from the outside faces of the rafters, you locate the center of each rafter by measuring over ¾ inch from each layout measurement or mark.

LAYING OUT THE RACKING

2. **Add the modules.** Sketch the module layout for the entire array onto your map. Maintain the required spacing at the ridge, eave, and side edges of the roof. Include all of the modules, but don't worry about making them perfect; you'll likely make some adjustments before finalizing the map. Refer to the racking specs for the spacing between adjacent modules in each row (determined by the module mid clamps) and the spacing between modules in different rows (typically ⅛ inch or more).

3. **Add the footers and rails.** Mark the footer locations on the map, centering each over a rafter. It's okay to vary the spacing to accommodate the actual framing layout, provided you don't exceed the maximum span between footers specified by the racking manufacturer, the building department, or engineer. Space the rows of footers evenly, according to the manufacturer's specs. Typically, the rails (which sit atop the footers) must be within the top/bottom 30% of the modules' length. There's also a maximum length that the rails and modules can extend beyond the outermost footers; this is typically 12 to 18 inches but varies by manufacturer. Draw the rails spanning across the footers. Depending on the type of module end clamps you use, rails may be flush with the outside edges of the outer modules, or they may need to extend 1½ inches or more beyond the modules to accommodate the end clamps.

LOCATING THE RAFTERS

Use the following steps to locate the rafters on your roof. If you have a vaulted, finished ceiling (so that the rafters are not visible from below), read For Finished Ceilings (opposite) along with the following steps.

1. **Drill a locator hole.** Your rafter measurements will start from a locator hole drilled through the roof at the top, outside corner of the racking layout. From inside the attic, identify the outer rafter that will receive footers. Measure down from the ridge board or peak of the roof frame and mark the approximate location of the first footer onto this first rafter. Drill a hole up through the roof sheathing and roofing, using a long ¼-inch (or smaller) drill bit, keeping the bit aligned with the **outside face** of the first rafter. Stick a straight piece of wire up through the hole and tape the wire to the side of the rafter. This will help you find the hole from the top side of the roof. Alternatively, you can have a helper mark the hole with a crayon.

2. Measure the rafters in the first (uppermost) row. Hook your tape measure on the **outside face** of the first rafter and measure straight across the rafters perpendicularly. Note the precise location of the **outside face** of each rafter you will anchor footers into, recording the dimensions on your rooftop map.

3. Mark the remaining rows. Back at the first rafter where you drilled the starter hole, measure down (toward the eave) along the length of the rafter and mark the locations of each of the remaining rows of footers, according to your plan. At the bottom row, drill another hole through the roofing and mark its location with another piece of wire (or have your helper mark it on the rooftop). When you're up on the roof, you'll snap chalk lines between the holes to create a vertical reference line. If the rafter bows (curves sideways) quite a bit, you may want to drill an additional hole for some of the intermediate rows. This first rafter is the reference point for all of your rafter measurements, and a bow of more than ½ inch or so can result in missing some rafters with the footers' lag screws.

4. Locate the remaining rafters. Measure across the rafters for each remaining row, noting the locations of the anchoring rafters, as before. It's necessary to measure each row because the rafters may not be perfectly parallel to one another and/or each rafter may not be straight.

TIP

FOR FINISHED CEILINGS

To locate rafters on vaulted ceilings, use a stud finder to locate each rafter and mark its side edges. Before drilling your starter hole, make sure there are no electrical wires or plumbing pipes in the area. Measure over from the starter hole to each anchor rafter; then confirm the rafter location with the stud finder. For more accuracy, drive a small finish nail through the drywall at ¼-inch intervals to pinpoint the rafter edges; you'll know you're on the rafter when you hit wood.

NO RAFTER WHERE YOU NEED ONE?

Sometimes the roof framing doesn't agree with your racking layout, due to inconsistent spacing, roof penetrations, or modifications to the framing. The standard solution is to add wood blocking between two rafters where you need an anchor point. Use framing lumber that is the same size as the rafters (2×6, 2×8, etc.), and cut the blocking to fit snugly between the rafter pairs. Tap the blocking into place and fasten it by driving two 3½-inch screws or two 16d nails through the rafters and into each end of the blocking. Alternatively, you can use 4×4 lumber for blocking, which gives you a little more wiggle room for anchor placement.

Tip: If you discover you need blocking while installing the module racking, drill a small pilot hole down through the top of the roof where you need the anchor point. Inside the attic, position the blocking so that it is centered under the hole.

MARKING THE ROOF

1. **Snap a vertical line.** On the rooftop, measure over ¾ inch from each starter hole of the array layout and make a mark, indicating the center of the rafter. With a helper, snap a chalk line through the marks to create a vertical (ridge to eave) reference line representing the center of the first anchor rafter. If you made additional holes to follow a bowed rafter, snap a chalk line between each adjacent pair of holes, like connect-the-dots.

2. **Snap the top horizontal line.** The horizontal lines mark the anchor locations for the footers. Measure down from the roof ridge at both ends of the layout and mark the height of the top row of footers. Snap a chalk line through the marks to create a horizontal reference line, perpendicular to the vertical chalk line.

NOTE: *Footers must be centered over rafters for proper anchoring, but often the rail locations can be moved up or down the rafters a bit so that the footers sit properly on the shingles (provided the modules don't overhang the rails by more than 30% of their length, or as specified). If necessary, adjust your horizontal line as needed so that the footer flashing will sit flat atop the shingles (see page 111).*

3. **Mark the remaining rows.** Measure down from the top horizontal line to mark the locations of the remaining rows of footers, measuring at both ends to make sure all lines are parallel. Adjust the row locations as needed to accommodate the shingles. Snap chalk lines through each set of marks.

4. **Mark the anchor points.** Using the dimensions from your rafter layout, measure over from the vertical chalk line and mark the location of each anchor point on the horizontal lines. Because you measured over ¾ inch from the starter holes to center the vertical line over the first rafter, you don't need to do that with the rest of them; just use the rafter measurements on your map, and all your marks will be centered over a rafter.

NOTE: *Doing this portion of the installation (measuring and marking) accurately is essential for the final PV array to end up looking straight and true, with all the rails and modules in perfect alignment.*

TIP

DRAW THE ARRAY CORNERS

Use a marking crayon to clearly mark the four corners of the array (that is, the outside corners of the outer modules) onto the roof. This will help you visualize the array's position until the modules are up. It can be an invaluable reminder, as you'll see the first time it prevents you or one of your crew from drilling in the wrong place. You can also mark the locations of the rails, but use a different color of chalk to distinguish the rail markings from the array markings.

Installing Footers and Rails

Tools

- **Drill**
- **Drill bit (sized for lag screws)**
- **Socket driver bit and socket (for lag screws)**
- **Caulking gun**
- **Flat pry bar or shingle ripper**
- **Ratchet wrench**
- **Torque wrench**
- **String line**
- **Wood blocks (2, must be the same thickness)**

Materials

- **Racking system components**
- **Lag screws (size and material specified by racking manufacturer/ building department/ engineer)**
- **Roofing sealant**

Installing the racking hardware itself often is the most product-specific aspect of the entire PV installation. Always follow the manufacturer's instructions for your hardware and roofing type. The steps shown here may or may not apply to your project, but the procedure demonstrates some of the essential basics – namely, preparing shingles for flashing, sealing around the penetrations, and torquing the bolts on the racking system.

A couple of specialty tools are required: a flat pry bar or a shingle ripper for removing nails from underneath shingles, and a torque wrench for making sure the fasteners are tight enough without overtightening. Also be sure to use the highest-quality roofing sealant you can find. Do not use caulk (silicone or other) or any other type of sealant not specifically designed for roofing. You can buy all of these supplies at local hardware stores and home centers.

REMOVING NAILS FROM SHINGLES

To remove a nail, slide a flat pry bar or shingle ripper (a $20 tool designed for this purpose) under the butt edge of the first shingle above the anchor point, breaking the adhesive seal where the shingles overlap. Slide the bar up under the nail (there will be a shingle between the bar and the nail head) and twist and pry the bar slightly to raise the nail head about ⅛ inch. Pull out the bar, slide it under the second shingle above the base, and pry up the nail directly. Remove the nail and cover the nail hole with a small amount of roofing sealant.

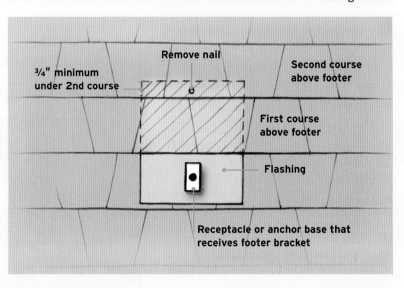

¾" minimum under 2nd course

Remove nail

Second course above footer

First course above footer

Flashing

Receptacle or anchor base that receives footer bracket

INSTALLING FOOTERS AND FLASHING

1. **Drill pilot holes.** Drill a pilot hole through the roofing and into the center of the rafter at each anchor point, using a bit that's sized for the specified lag screw. A common size for the lag is ⁵⁄₁₆ inch, which typically calls for a ³⁄₁₆-inch pilot hole.

2. **Break the shingle seal.** Asphalt shingles have an adhesive strip underneath their butt edges that sticks to the shingles below. When the roof is hot, the tarlike adhesive is really soft and you can usually just slide the flashing through it. Otherwise, carefully break the seal with a flat pry bar. The flashing must slide up under two courses of shingles above the anchor point. Remove any nails in the way of the flashing, as needed (see Removing Nails from Shingles, opposite). Repeat this step as you install each piece of flashing (step 3).

3. **Install the flashing.** Fill the pilot hole (for the footer lag bolt) with roofing sealant. Install the footer flashing as directed by the manufacturer. You may need to apply roofing sealant to one or both sides of the flashing. Typically, the flashing slides up under the two courses of shingles above the footer base, and its bottom edge is near or slightly above the butt edge of the shingle it rests on. **Tip:** Use a screw as a guide to align the hole in the flashing with the pilot hole.

4. **Mount the footers.** Position each footer and footer base (as applicable) over a flashing piece and secure them with a lag screw through the hole in the flashing and into the pilot hole in the roof. If the footer can be adjusted on its base, leave it a little loose so you can move it as needed when positioning the rails (you will tighten the lag screws later).

Note: *Lag screws must be long enough to penetrate the rafters at least the minimum distance required by the racking manufacturer and local building department. Often the structural engineer (if you use one) will specify the size of lag screw required.*

ASSEMBLING AND ADJUSTING THE RAILS

1. **Attach the rails.** Attach the first section of rail onto the footers as directed by the manufacturer. Leave the rail connections loose for now so you can adjust the rails later. If the row is longer than a single rail, attach the second rail to the footers, then splice the rails together using the provided splicing plate or clamp. Be sure to leave the specified gap between rail sections to allow for thermal expansion (¼ inch is typical). Tighten the lag screws on all of the footers in the row. Repeat the same process to assemble all the rails in every row.

TIP

BOTTOM UP

Most PV installation pros install the rails and modules from the bottom up, starting near the eave and working toward the ridge. The rails make for handy supports for your feet, your tools, or even a module or two. Just don't stack a bunch of modules on the roof, creating the potential for one really expensive avalanche!

2. **Set the rail height at the ends.** Starting at the outside footer at one end of the bottom rail, position the rail so its mounting bolt is roughly centered top-to-bottom in the elongated hole in the footer (the elongated hole allows for adjustment up or down). Tighten the bolt to the specified torque setting. Repeat with the last footer at the other end of the rail.

NOTE: *The rails — and ultimately the modules — are installed parallel to the roof and in a flat plane. They do not have to be level (perfectly horizontal), like a countertop or a pool table. Roofs are not always level from side to side. If the roof is out of level and the modules follow it, they will look fine. If the roof isn't level but the modules are, you might see the difference from the ground.*

3. **Set up a string line.** Tie a string line onto the lag bolt of one of the outside footers. Pull the line around the end of the rail and over the top. Run the line to the other end of the rail and tie it off in the same way, making sure it is taut. Slip a block of wood under the line at each end (blocks must be the same thickness; as shown, the blocks are ¾ inch thick) so the string is held above the rails.

4. **Straighten the rail.** Working from one end to the other, measure between the string line and the top of the rail. Adjust the rail height at each footer so the rail is ¾ inch (or whatever your block thickness is) from the line, then tighten the rail mounting bolt to secure it to the footer, torquing the bolt as directed. Repeat at each footer until you reach the other end of the rail.

5. **Straighten the remaining rails.** If there is only one row in the array (two rails total), move the string line to the upper rail and repeat the same process to straighten and secure the upper rail. If there are more rows in the array, straighten and secure the top-most rail as you did the bottom rail, and then tie the string line to the top and bottom rails so it runs perpendicularly over all of the rails. Use the same measuring technique to position the intermediate rails (those between the top and bottom rails), so they are the same distance from the string line as the top and bottom rail. This ensures that all of the rails are in the same plane.

6. **Check all fasteners.** Double-check all of the racking connections (footers and rails) before moving on to the module installation. Start at one end of the top or bottom rail, and move down each rail or row, retorquing all the nuts/bolts with the torque wrench.

Installing Modules

Once the footers and rails are in place, aligned, and secured, installing modules is a straightforward process of assembling prefabricated parts. The steps and techniques vary by the type of system and the specific products used. The general procedure involves four tasks:

- Mounting microinverters or DC optimizers (as applicable) to the racking rails or to the module frames (see page 116). With some microinverter systems, this step may include securing AC trunk cables along the rails (the microinverters then connect to the trunk cables). With string-inverter systems, it may be easiest to install the long home-run cables along the racking before installing the modules.

- Grounding (or bonding) modules, rails, and microinverters/DC optimizers to a system ground. While module clamps and WEEBs typically bond modules to rails, bonding the rails requires a separate ground wire. Microinverters may or may not need to be attached to this ground wire, depending on the model and the system wiring.

- Installing a junction/combiner box (which may be either DC or AC).

- Installing modules, wiring them together as you go and securing them to the rails with mid and end clamps.

The system shown in the following steps includes microinverters that serve two modules (rather than one) and connect together with their own wire leads; they do not connect to a separate AC trunk cable. The microinverters and rails are grounded/bonded with a single ground wire. Wiring details for string inverters are discussed in Home Runs for String-Inverter Systems, on page 120.

Tools

- **Ratchet wrench and sockets**
- **Torque wrench**
- **Wire cutters**
- **4- or 6-foot level (or straightedge)**
- **Jigsaw**
- **Needle-nose pliers**

Materials

- **Grounding lugs**
- **6 AWG solid-copper ground wire (or as specified by local code)**
- **Black UV-resistant wire ties or outdoor cable clips**
- **PV modules**
- **Module end and mid clamps**
- **WEEBs (as needed)**
- **Rail caps (as applicable)**
- **Array screening system (optional)**

Additional Materials for Microinverter Systems

- **NEMA 3R electrical AC junction box**
- **AC trunk cable (one per branch circuit, as applicable)**
- **Microinverters**

Additional Materials for String-Inverter Systems

- **Disconnecting DC combiner box**
- **10 AWG PV cables (one two-ended cable per series-string, for home runs)**
- **DC optimizers (as applicable)**
- **Electrical tape**
- **Marker**

MOUNTING MICROINVERTERS (OR DC OPTIMIZERS) TO MODULES

Installing microinverters or DC optimizers to the backsides of modules is called frame mounting and is an alternative to rail mounting. Frame mounting, available with most models, is primarily used with railless racking systems and ballasted (flat-roof) systems, which don't have rails to mount to, but it can also be used with conventional rail systems. It typically requires a separate mounting bracket supplied by the inverter or optimizer manufacturer.

To frame-mount a microinverter or DC optimizer, choose a location where the unit will not interfere with the rail when the module is installed. Secure the unit to the module frame and torque the fasteners as specified. Connect the wire leads from the module to the microinverter or DC optimizer as directed. When you install the modules on the rooftop, you will connect the outgoing leads from the microinverters to one another or to the AC trunk cable; optimizers also will connect to one another.

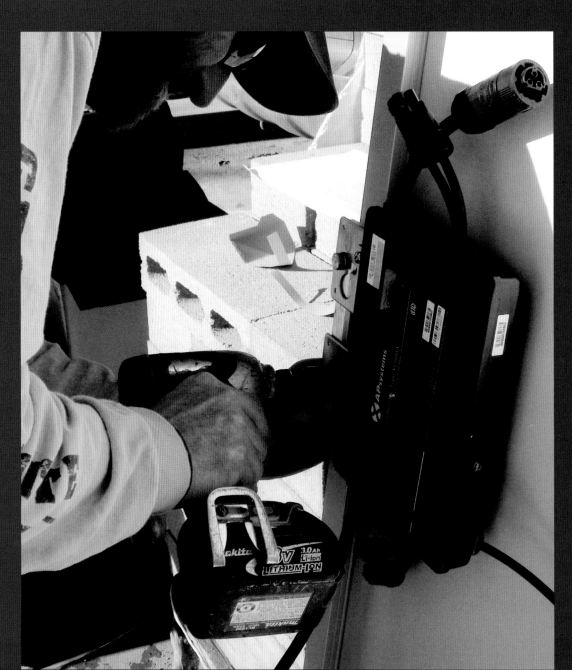

INSTALLING THE MICROINVERTERS AND GROUND WIRE

1. Mount the microinverters to the rails.
Mark where the centers of the modules will fall on each of the rails that will receive microinverters (every other rail, since the units mount to only one rail for each row of modules). Install a microinverter at each mark, following the manufacturer's directions. Usually this involves one or two brackets with holes and bolts that fasten the microinverter bracket to the rail.

2. Connect the microinverters together.
Join each microinverter to the next one in the row using the connectors on their wire leads. Alternatively, if the system includes an AC trunk cable, extend each trunk cable along the rails with the microinverters. Align the cable connectors on the trunk cables with the microinverters so that the inverter leads can reach the connectors. Plug in each inverter lead to a connector, as directed by the manufacturer.

3. Cap off the last microinverter lead.
The last microinverter in each row or branch circuit will have an unused lead, since there's no other inverter down the line to connect to. Cover this lead with a termination cap supplied by the manufacturer. Most simply fit over the lead connector and twist to lock. The first module in the row or branch circuit will connect to the junction box with a special cable that has a module connector at one end and loose wires at the other end. AC trunk cables have similar termination caps; install these as directed by the manufacturer.

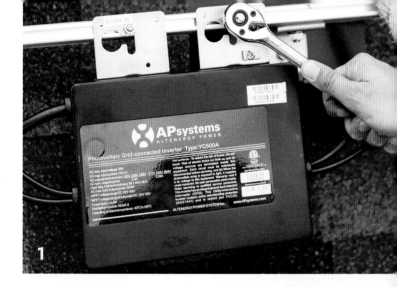

4. **Mount the grounding lugs.** Install a grounding lug on each rail, and torque as directed by the manufacturer. Lugs are typically arranged in a straight line across and perpendicular to the rails – to save on wiring – but can be installed anywhere on each rail, as desired.

5. **Install the ground wire.** Secure the 6 AWG solid-copper ground wire to the first lug (typically on the rail farthest from where the array's junction/combiner box will be). Tighten the lug over the wire to the specified torque setting. Extend the wire in a straight line to the next connection point. The ground wire **must** connect to all of the rails. **Tip:** Make 90-degree turns, as needed, for a neat installation. In the lower left photo, a little pigtail spiral (made with pliers after the wire is secured) is one installer's finishing touch at the start of the wire run.

6. Ground the microinverters (as applicable). Connect the ground wire to each microinverter case or mounting bracket, as directed. Run the wire to the junction/combiner box location, leaving enough slack for wiring into the box later.

NOTE: *As mentioned on page 115, some microinverters are designed with internal grounding and do not need to use the external ground wire. However, the system must include a ground wire in all of the cabling between the microinverters and into the junction box. If any of this cabling lacks a ground, you'll need to connect the microinverters to the external ground wire.*

7. Tie up the wiring. Secure the cables to the rails with cable clips or wire ties (see Wire Management, page 123).

8. Install the junction box. Mount a NEMA 3R (outdoor-rated) electrical box of a suitable size to the racking rail or to a conduit located near the rail (see Rooftop Conduit Runs, page 137). **Note:** Use only the pre-drilled holes and/or mounting tabs for fastening the box to the rail or conduit; do not drill new holes or otherwise alter the box. **If you have a string inverter system,** you will install a disconnecting combiner box instead of a junction box. This must be within 10 feet of the array to satisfy the NEC's rapid shutdown requirement.

With microinverters, you will be bringing AC into the junction box. With string inverters, you will be bringing DC into the disconnecting combiner box.

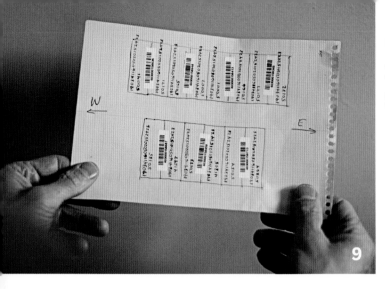

9. **Map the installation.** Create a simple map or chart of the entire array, noting the location and serial number of each micro-inverter and module. Most microinverters come with stickers that you can peel off and add to your map. This record will prove invaluable if an inverter has a problem and you need to know where it is located in the array. Also, your inspector and/or the utility may request to see the map.

HOME RUNS FOR STRING-INVERTER SYSTEMS

Each series-string in the array gets two home run cables: one runs from the last module in the string and along a rail back to the combiner box; the other (typically shorter) runs from the first module in the string to the combiner box.

NOTE: If you're adding DC optimizers to your system, you will install the optimizer units to the rails (similar to microinverter installation) and connect the units to one another. The modules connect to the optimizers. The home runs for the strings connect to the first and last optimizers in the strings.

An easy way to make home runs is to use long PV cables with factory-installed quick-connect fittings on both ends. (The fittings are typically MC4 connectors but must be compatible with your modules.) Extend each cable from the far end of each string back to the combiner box. Cut off the end at the combiner box and use the leftover cable, with its connector, to make the shorter run from the near end of the string to the combiner box. Plan carefully so that you have the proper fittings at the ends of the strings: if the last module will have a female lead for connecting to the home run, start that home run with a male fitting, and vice versa.

Label the ends of the home runs at the combiner box, using light-colored electrical tape and a marker. Assign a number to each series-string, and note the string number and whether the cable is connected to a positive (+) or negative (−) module lead. Use a tape color that is not red or green. Red usually means positive (+), and green usually indicates ground, in standard electrical installations.

WARNING: Leave the home run cables disconnected until the end of the PV system installation. **Do not connect the home run cables to the modules** when the modules are installed. Modules create electricity under any daylight conditions. Leaving these cables disconnected is the best way to ensure safety until the system installation is complete.

Note: With microinverter systems, it's okay that the modules and microinverters are connected to the AC trunk cable, because the inverters don't output electricity unless the PV system is connected to the utility grid.

INSTALLING THE MODULES

1. **Position the first module.** Set the first module onto the rails at one end of the row (if the array has multiple rows, start at the bottom row). Fit two end and mid clamps onto the rails, using WEEB washers, if necessary (see Grounding Your Modules and Racking, page 43). Measure to confirm that the module is centered top-to-bottom and is square to the rails. Connect the module leads to the rail-mounted microinverter or DC optimizer, if applicable. If you have frame-mounted microinverters, connect the microinverter to the AC trunk cable, as applicable.

2. **Position the second module.** Set the second module into place and snug it up against the first module with the mid clamps (with WEEBs, as applicable) in between. Align the bottom edges of the first two modules, using a 4- or 6-foot level (simply as a straightedge, not as a level). Secure both modules with the mid clamps, and torque the clamp bolts as specified by the manufacturer. Connect the wiring as applicable.

NOTE: *Since WEEBs are installed two for every two modules, you will place the WEEBs under every* other *pair of mid clamps, and under the end clamps on rows with an* odd *number of modules.*

- Microinverter (rail-mounted): Connect second module leads to microinverter (microinverters already connected together or to trunk cable).

- Microinverter (frame-mounted): Connect microinverters together or to AC trunk cable (modules already connected to microinverters).

- String inverter without DC optimizers: Connect first module + lead to second module – lead, or vice versa (remember you are wiring them in series).

- String inverter with rail-mounted DC optimizers: Connect each + and – module lead to its own optimizer unit (optimizers are already connected together).

- String inverter with frame-mounted DC optimizers: Connect first module optimizer to second module optimizer (modules are already connected to optimizers).

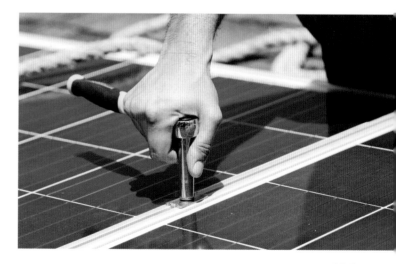

3. **Continue down the row.** Install more modules in the row, using the same techniques and adding WEEBs as needed. Check the alignment of each module with the level. When the row is complete, it's a good idea to view the entire row from various angles and from the ground to make sure the modules are visually aligned with the roof. Visual alignment, as seen from the ground, is more important than precise alignment and squareness to the rails.

4. **Complete the row.** Secure the outside module at each end of the row with a pair of end clamps, using WEEBs, as applicable. Retorque all of the clamps in the row. **Note:** Remember to call for your rough inspection at the specified time, based on what the inspector wants to see of the modules and racking.

WARNING: Do not connect the home-run cables to the modules (string-inverter systems). Leave them disconnected until the end of the PV system installation, when they can be connected by, or under the guidance of, your electrician. Modules create electricity under any daylight conditions **(a series-string of connected modules produces a dangerously high voltage)**. Leaving these cables disconnected is the best way to ensure safety until the system installation is complete.

5. **Install the remaining modules (as applicable).** If there are multiple rows in the array, install the modules in the next row up from the bottom, using the same techniques as before. Maintain the manufacturer's minimum specified gap between the rows (typically ⅛ inch or more). Sight down the row as you go to keep the modules aligned. Complete the remaining rows, then trim the rails to length, as specified, using a jigsaw or reciprocating saw. Cover the ends of the rails with caps from the manufacturer, if applicable. Add screening to enclose the space below the modules to keep out animals, if desired (see page 41). Sometimes the ends of the rails are spray-painted to match the color of the module frames.

WIRE MANAGEMENT

Neatness is one of the hallmarks of quality on any electrical job, but it's particularly critical for PV systems. Dangling wires look bad — even if you can't see them from the ground, you'll know they're there, and your inspector might frown on them. They're also prone to collecting dirt and debris and are vulnerable to damage from sliding snow and ice. Make sure that wires do not dangle down from the modules or rails and do not touch the roof. As a good rule of thumb, you should be able to look under your array and see no wiring below the bottom edges of the module rails.

To help tame and protect wires, some racking systems have rails with channels for laying in the wires. Rail manufacturers also offer cap pieces for enclosing the channels if there's a large gap between modules. If your rails don't have channels for wire management, secure the wiring to the rails with outdoor-rated wire clips or black, UV-resistant plastic wire ties (zip ties).

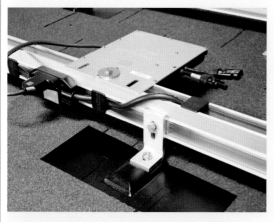

Rails with channels offer the cleanest installation. The channels are open at the top but are almost entirely covered by the modules.

Plastic wire ties can wrap all the way around rails and are handy for tying up coils of wire.

Rail clips (typically supplied by racking or microinverter manufacturer) fasten directly to rails.

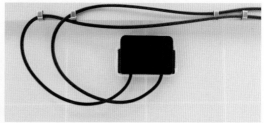

Wire clips for module frames help tame long module leads. These are usually metal or plastic, ¼ inch in size.

INSTALLING FLAT-ROOF SYSTEMS

Ballasted flat-roof systems offer a nice advantage over sloped-roof systems: you don't have to locate and anchor into rafters. Most installations follow an assemble-as-you-go process. Here's an overview of the basic steps (installation details can vary widely from system to system):

1. Snap a chalk line onto the roof to represent the front edge of the first row. Position two front module brackets on the line. Position two brackets for the rear of the module, spacing them according to the module size. If there will be multiple rows in the array, use connector-type rear brackets. If there will be one row only, use standard rear brackets. Weight the brackets with one ballast block (concrete block) each, as applicable. **Note:** Some systems include metal trays for ballast blocks; others may require protective material between the blocks and the roof.

2. Set the first module onto the brackets, and adjust the brackets as needed. Add module end clamps and mid clamps to the brackets.

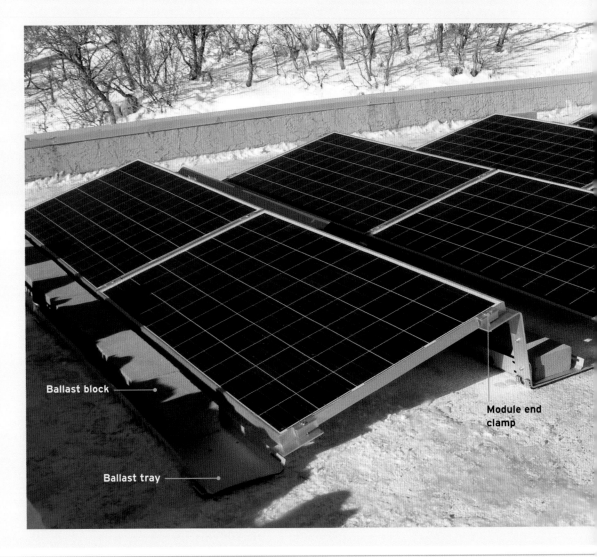

Ballast block

Ballast tray

Module end clamp

NOTE: Since there are no rails, micro-inverters and DC optimizers are frame-mounted to the modules prior to installing the modules (see page 116).

3. Position and weight the next two brackets; then add the second module. Secure both modules with the module clamps. Connect the modules/microinverters/

DC optimizers together electrically as required for the system type. Continue the same process to complete the first row.

WARNING: Do not connect the home run cables to the modules for string-inverter systems (see WARNING, page 122).

4. Begin the next row behind the first. The spacing between rows is automatically set by the connector brackets. Complete the remaining rows in the array. For the last row, use standard rear brackets (not connector brackets) to support the rear edges of the modules.

5. Complete the ground wire and AC/DC prewiring runs per the system design.

6. Install wind deflector panels to the rear and connector brackets, as applicable. Add ballast blocks as needed to meet the design specifications. Specs should include the quantity, weight, and location of ballast blocks for the given tilt and wind and snow loads.

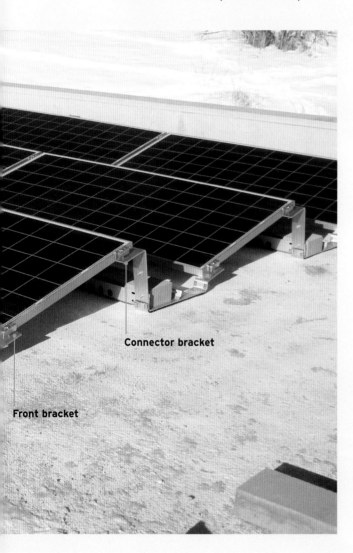

Connector bracket

Front bracket

7 Mechanical Installation:
Ground-Mount

GOAL

Build the ground-mount structure and install the PV modules, microinverters or DC optimizers (as applicable), and grounding wires to complete the array

CHAPTER 6 COVERED THE INSTALLATION of the entire PV array for rooftop systems. This chapter does the same for ground-mounts, showing you how things differ for a system on terra firma. Once the module rails are installed on the main ground support structure, you'll refer to chapter 6 for the module installation (and prewiring), which will be the same as for rooftop systems. Much of what happens at this stage is specific to your equipment, your module orientation, and your installation site. Be sure to follow the manufacturer's instructions and the local building code for every aspect of your project. We'll go over the basic process and guide you through the trickiest part of most installations – setting the posts – but the construction details and sequence of steps must come from the manufacturer's instructions and plan drawings. The main stages of the process are setting the posts in concrete, assembling the support racking, and adding the modules. Let's break ground!

Setting the Support Posts

Tools

- **Circular saw**
- **Hammer or drill**
- **Sledgehammer**
- **Line level**
- **String line**
- **Digging tools (shovel, posthole digger, power auger, digging bar, as desired)**
- **Marker**
- **Level**
- **Camera**
- **Reciprocating saw or chop saw**
- **Concrete mixing tool**
- **Masonry trowel**

Materials

- **2×4 lumber (6 @ 8 feet each)**
- **2½-inch nails or wood screws**
- **Masking tape**
- **Schedule 40 galvanized water pipe (or other material, as specified by manufacturer or engineer)**
- **Wood stakes**
- **1×2 or 1×4 lumber**
- **1½-inch wood screws**
- **Duct tape**
- **Gravel**
- **Concrete mix**

Installing the vertical posts for a standard ground-mount structure is a lot like setting fence posts or laying out concrete piers for a deck project. A reliable (and inexpensive) method is to use string lines for laying out the holes and positioning the posts. String lines are most useful when they're tied to batter boards, simple assemblies of two vertical wooden stakes with a horizontal cross-piece. Once you've set up a pair of batter boards at each corner of the installation area, you tie on the strings and get them level and square. The strings will guide your hole and post placement, and the batter boards allow you to remove and reattach the strings as needed without having to recheck them for level or square.

This sample project is for two rows of vertical posts – a row of short posts in front and a row of long posts in the back. The project steps follow the standard process for installing the posts in concrete before assembling the rest of the ground-mount structure. An alternative method that's an option with some systems is to preassemble the main support structure before concreting the posts. This makes the post positioning more foolproof and often is a good option when the structure is a manageable size, but overall it requires about the same amount of work as the conventional method.

WARNING: Have your utility lines marked before you do any digging. Call 811 to notify the national "Call Before You Dig" hotline. An operator will notify all utility providers in your area. Any company that has a buried line on your property is required to come out and mark it within a specified period, typically two to three days. It's that easy, and it's all free. It also could save you from a deadly encounter with a gas or electrical line or a deadly-smelling run-in with a waste pipe – in case you needed more incentive to call.

1. **Build the batter boards.** Construct eight batter boards with 2×4 lumber and 2½-inch nails or screws. Cut the vertical legs to length at 24 inches, trimming one end to a point. Fasten a 24-inch crosspiece across the tops of the legs so that the legs are parallel and the crosspiece is perpendicular to the legs.

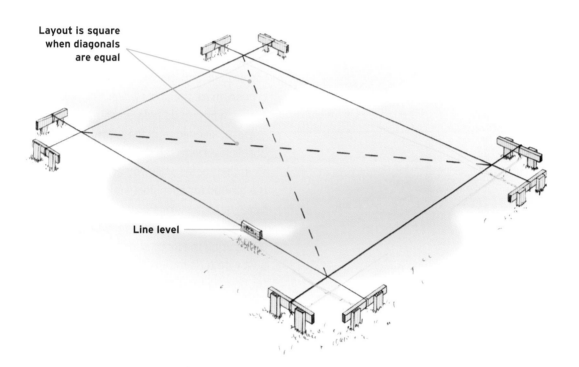

Layout is square when diagonals are equal

Line level

2. **Set up the string lines.** Drive a pair of batter boards into the ground, at a right angle to each other, at each corner of the installation area. Tie four lengths of string line to the batter boards to form a rectangle. The strings should cross about 2 feet away from the batter boards. Square the lines by measuring diagonally between opposing corners, moving the strings as needed until the diagonal measurements are the same. Level each string with a line level, tapping in one of the batter boards as needed to adjust. Mark the position of each string onto the crosspiece it is tied to; this shows you where to retie the strings without losing the square layout.

3. **Dig the post holes.** Once the strings are square and level, measure from the corners of the string layout, and mark each post location with masking tape and a marker, following the manufacturer's drawings. Using an auger or post-hole digger, dig the post holes directly below the marks. If desired, you can untie the strings at one end so they're not in the way. Dig the holes to the diameter and depth specified by the ground-mount manufacturer and the local building code or project engineer. The holes must extend below the frost line for your area. It's also a good idea to add 4 to 6 inches of depth for drainage gravel.

4. **Set and brace the posts.** Add 4 to 6 inches of gravel to each hole and compact it with a scrap 2×4 or a hand tamp. Retie the layout strings, if necessary. Cut each post to length with a reciprocating saw or chop saw. Mark the posts at the required height relative to the string line. For example, if the posts should stand 3 feet above the ground, and the string line is 1 foot above the ground, mark the posts 2 feet from the top ends. Set each post in its hole, and add or remove gravel as needed until the height mark on the post is level with the string line. Let the posts sit in their holes while you drive two stakes in the ground for each post and screw a 1×2 or 1×4 brace to each stake. Position each post in the hole, plumb it with a level, and brace it in place by securing the loose ends of the braces to the post with duct tape. It helps to have an assistant for bracing the posts.

TIP

PHOTO EVIDENCE

When the holes are dug, take a photo of your tape measure sticking out of the hole; the inspector might want proof of the hole depth. If the top of the footing won't be visible, also photograph the tape going across the hole, to show its diameter.

5. **Add the concrete.** Fill the post holes with concrete. You can mix bags of concrete in a wheelbarrow or a mixing tub. For large projects, it might be worth it to order ready-mix concrete from a local supplier (call suppliers for details; they may have a minimum order size or charge extra for partial yards of concrete). You can fill the holes so that the concrete is flush with or slightly above ground level, or leave it about 4 inches below ground level if you want to grow grass over it. Use a trowel to slope the top of the concrete away from the post to promote drainage. Let the concrete cure for three to four days, depending on weather; then remove the bracing.

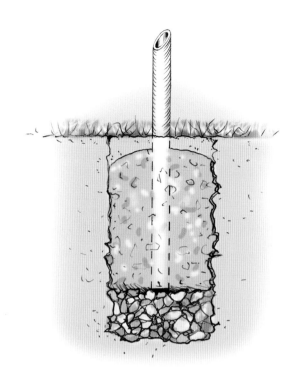

ANCHORING POLE MOUNTS

Single- and multiple-pole ground-mount systems have relatively few anchor points, so they typically require larger support poles and larger concrete foundations than conventional ground-mounts. The basic steps for installing the posts and concrete are similar, but large concrete foundations may need internal reinforcement (such as rebar structures), and you may want to embed conduit in the concrete so that you can run wiring through the foundation, rather than having the conduit exposed on top. It also may be permissible to install the grounding rod (before the concrete is poured) so it comes up through the foundation.

Foundations for pole mounts usually must be designed by an engineer for approval by the building department. Structures for large arrays are best left to professionals.

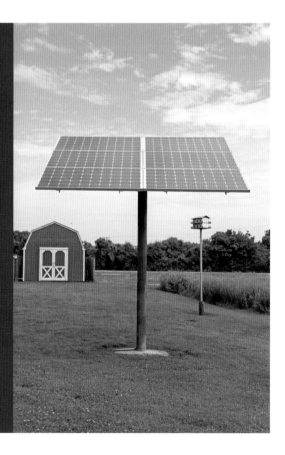

Assembling the Ground-Mount Structure

Tools

- **Ratchet wrench**
- **Torque wrench**
- **Reciprocating saw or chop saw**
- **Metal file**
- **Wire strippers**

Materials

- **Ground-mount system rails and hardware**
- **Ground wire and lugs**
- **Modules and module clamps**
- **Combiner or junction box**
- **Home run wiring/branch circuit cables**

The typical assembly process starts with mounting base rails across the vertical posts. These are followed by the module rails and, finally, the modules. Some systems also include cross-bracing between the front and rear rows of vertical posts and/or between posts in the same row.

The structure in the steps shown here is designed for an array with **landscape orientation**. This means that the base rails are horizontal and level and run parallel along the tops of the front and rear rows of posts. If your modules will be in **portrait orientation**, the base rails will run perpendicular to the post rows and will angle upward from the front to rear posts, as shown below.

1. **Install the base rails.** Fit the post brackets onto the posts and secure them as directed by the manufacturer. Depending on the system, you might want to leave the brackets loose for now, to allow for adjustment. Mount the base rails to the post brackets, again keeping them a little loose, if applicable. Join lengths of base rail as needed, using the system hardware (or pipe couplings, if the rails are standard pipe). Plan the location of the couplings so they won't interfere with the post brackets or the module rail brackets. Trim pipes to length with a reciprocating saw or chop saw. File the cut ends smooth (you can add caps later, if desired). When all the supports are in place, confirm that everything fits properly and is square and level; then tighten the hardware/fasteners and torque them as directed.

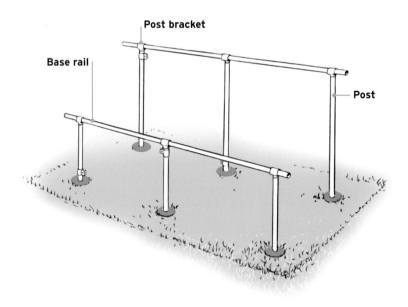

Post bracket

Base rail

Post

2. Mount the module rails. Mount the module rails to the base rails, using the system hardware. Space the rails according to the module specifications, and plan for the required gapping between modules in the same row (typically set by the module mid clamp) and between modules in adjacent rows (typically ⅛ inch). Install bracing members between the posts per the structure design and engineering requirements.

3. Prewire the array and install the modules. The steps for this part of the installation are the same as for a rooftop system and are covered in Installing Modules (page 115). With the ground-mount structure, you have the option of mounting your junction box or combiner box to the structure, rather than the module rail. You can also install horizontal pieces of metal framing channel (Unistrut is one brand) across two posts and mount the boxes to the channel. It's also possible to mount a string inverter to this channel, provided it is rated for outdoor exposure. It should be protected by the array as much as possible.

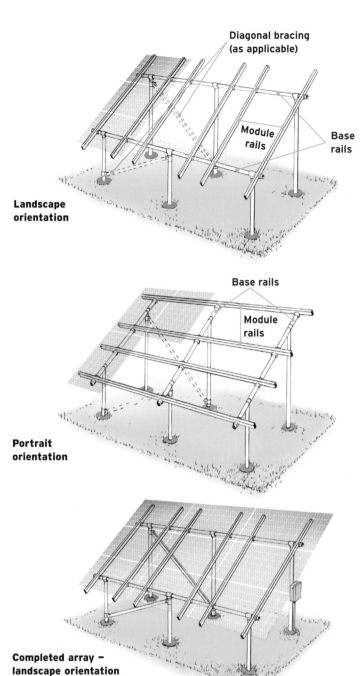

Diagonal bracing (as applicable)

Module rails

Base rails

Landscape orientation

Base rails

Module rails

Portrait orientation

Completed array — landscape orientation

RAPID SHUTDOWN RULES FOR GROUND-MOUNT ARRAYS

If you have a string-inverter PV system and you're supplying solar-generated power to your house, you almost certainly need rapid-shutdown protection. Technically, rapid-shutdown protection is required only for the parts of the system that are on or in your house (or any type of building); the outdoor parts of the system that do not touch the house do not need protection. But in most cases the DC combiner-disconnect box will be mounted next to the ground-mount array. Therefore, the rapid-shutdown system will be capable of cutting power to all of the wiring between the array and the DC–AC inverter at the house.

As with rooftop PV systems with microinverters, a separate rapid-shutdown system is not required for ground-mount arrays with microinverters.

8 Electrical Installation

GOAL

Complete the final installation phase of your grid-tied PV system

THIS CHAPTER PICKS UP WHERE WE LEFT OFF with the mechanical installation for rooftop and ground-mount arrays and covers the rest of the installation of a grid-tied system. The transition from the largely mechanical tasks to the final electrical work is a natural place to stop and decide how much more you want to do yourself. For most DIYers, this is a good time to turn everything over to an electrician (and, ultimately, a utility worker). In fact, this is when many professional solar installers bring in their own electricians or electrical subcontractors. This is because the remaining jobs require expertise with electrical systems and materials, and in most jurisdictions all AC electrical equipment and wiring *must* be installed by a master electrician (not a journeyman). If you know how to bend conduit and pull wires, you might decide to take care of these jobs before having your electrician make the final connections. Either way, conventional wisdom (and our recommendation) is to have a professional install and connect everything beyond the DC–AC inverter as well as the rapid-shutdown equipment. We'll discuss what's involved so you'll understand all the parts of your system and will know what to expect from the professionals.

Final Project Steps

With the rooftop or ground-mount array completed, the next step is to run a conduit down to ground level (from a rooftop array) or over to the house (from a ground-mount array), where it will tie into the remaining components and the utility grid. Here's what's left for the whole project, in a nutshell:

INSTALLING CONDUIT AND WIRING between the combiner or junction box at the array and the system components at ground level. This job entails the last big chunk of grunt work. You can save some money by doing some or all of the work yourself, if you're comfortable with the techniques. In any case, it's a good idea to discuss your plan with your electrician to ensure that all of your work will meet his or her standards.

INSTALLING AND HOOKING UP THE GROUND-LEVEL COMPONENTS, including the following, depending on your system type (we also note who should install them):

◇ DC–AC inverter: string-inverter systems only; you can mount the inverter to the wall (it's easiest if the inverter comes with a mounting plate), if you like, but let the electrician make the wiring connections

◇ Rapid-shutdown control: string-inverter systems only; electrician install

◇ AC disconnect: electrician install; must be labeled for identification

◇ PV production meter: applies to all systems but is not always required by your utility; electrician install; must be labeled for identification

◇ AC breaker(s) for PV power in main electrical service panel: electrician install; must be labeled for identification

◇ Utility net meter: installed by the utility (or one of their subcontractors), generally at no cost, but it might take them a while to get to it; usually bears a sticker with the word *NET*

WARNING: The *net* meter allows the utility meter to run in both the forward and backward direction. Do NOT turn on your PV system without a net meter and utility approval.

Running Conduit

All aboveground wiring linking the array and all the other system components must be enclosed in electrical conduit. Electrical metallic tubing (EMT) conduit is the industry standard, although some inspectors may approve flexible metal or polyvinyl chloride (PVC) conduit in certain situations. But you can't go wrong with EMT. Requirements and options for running conduit depend on the type of system.

Rooftop Conduit Runs

With a rooftop array, you have the option of running conduit through the roof or out and around the edge of the roof. Going through the roof offers a much neater installation, since there's no conduit running across the roof and wrapping around the edge. Instead, the conduit passes through the roof (through an approved boot with flashing) and into the attic, where it's routed toward an exterior wall, then through a soffit at the eave and down the outdoor wall (conduit may also run through a chase from the attic to an interior electrical room). This is usually the better option if you have an open attic with access to the underside of the roof.

Conduit installed over the roof should be anchored with flashed fittings. In some areas, it may be required that the conduit anchors elevate the conduit above the roof surface. This helps lower the temperature inside the conduit to prevent overheating and also allows rain and snow to slide underneath.

Conduit mounts with elevated anchors and flashing ensure a watertight installation and provide for cooling airflow underneath the conduit, similar to that underneath the PV modules themselves.

Penetrations through the roof require a flashed fitting, such as a boot (similar to a plumbing vent pipe penetration) or a sealed box enclosure, which doubles as a junction box for wiring transitions. Some box enclosures can be mounted under the array so that the box and conduit aren't visible from the ground. Flashed fittings are installed much like flashing for racking footers, with the flashing slipped up under the shingles above and sealed with roofing sealing as specified by the manufacturer.

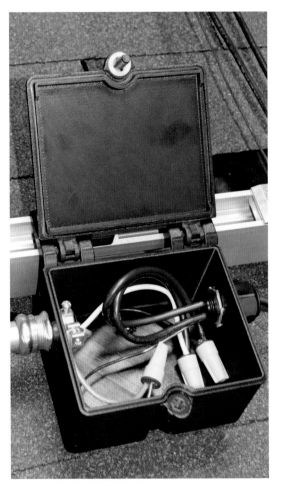

Standard penetration fittings (left) include metal flashing and a rubber boot that seals around the conduit. High-quality fittings have a seamless metal cone at the base so that the rubber seal sits above the roof level to reduce wear.

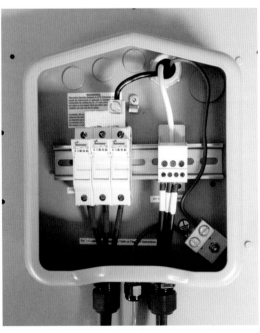

Box enclosures can function as junction or combiner boxes, in addition to anchoring conduit and making a watertight roof penetration. Locating a box under the array makes for a very clean installation; just be aware that you have to remove a module (or leave one off during the installation) to access the box.

Ground-Mount Wiring Runs

Wiring that connects a ground-mount array to the house typically runs underground and must be installed according to local code. In many areas, you're allowed to use underground feeder (UF) or underground service entrance (USE) cable, enclosing it in conduit only at the ends of the trench and wherever the cable runs aboveground. If you prefer to use conduit underground, options include schedule 80 PVC conduit and rigid steel conduit (RSC). PVC is inexpensive and easy to work with, while RSC requires threaded pipe and fittings as well as special tools for threading custom lengths of pipe. Running conduit all the way often means you can use THWN-2 wiring instead of UF cable.

COMMON REQUIREMENTS FOR UNDERGROUND WIRING

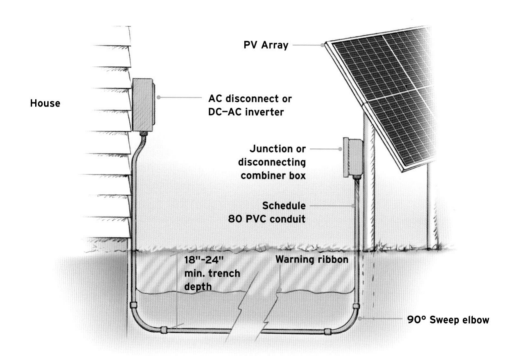

PV Array

House

AC disconnect or
DC–AC inverter

Junction or
disconnecting
combiner box

Schedule
80 PVC conduit

18"-24"
min. trench
depth

Warning ribbon

90° Sweep elbow

TIP

RUNNING CONDUIT ABOVEGROUND

You may be allowed to run conduit along or under a deck or other suitable structure to reduce the amount of trenching required. Be sure to gain approval from your building department before planning on this option.

Trenches for cable typically must be at least 18 to 24 inches deep and usually below the frost line. At each end of the trench — at the array and the house — the cable should enter the conduit through a 90-degree sweep elbow to ensure a smooth transition to the vertical conduit extending aboveground. Trenches for continuous PVC conduit don't need to be as deep — typically 18 inches. Burying a warning ribbon in the trench is a good idea for all installations and may be required by local code. Ribbons are usually buried 12 inches above the cable or conduit. If your inspector doesn't need to see the open trench, it's a good idea to take a photo of your tape measure extended to the bottom of the trench, just in case you need to prove that your trench is deep enough.

Communication cables (for data monitoring, rapid shutdown systems, and so forth), such as CAT-5 cable, typically must be run in a separate conduit from current-carrying wires.

Digging the trench yourself is a good way to save on the electrician's bill. The easiest method is to rent a trencher, a gas-powered, single-operator machine that has what looks like a giant chainsaw bar and teeth on the business end. Trenchers can dig 4- or 6-inch-wide trenches; 4-inch is sufficient in most cases. Whatever digging tool you use, call 811 before you dig (see WARNING, page 128) to prevent hitting utility lines.

Should You Pull Any Wires?

Pulling the wires between the array and the ground-level components might make sense if you know what you're doing and you've confirmed the wiring materials with your electrician. But this isn't a good project for beginners to cut their teeth on. The primary concern with pulling wires is that if you don't do it properly you can damage the wiring insulation and end up with an exposed conductor, creating the potential for a dangerous short in the circuit (exposed wires, metal conduit . . . you get the picture). Including plenty of pulling points in the conduit run makes it easier to pull wires and reduces the chance of damage. As for the wiring between the ground-level components, leave all of that to your electrician.

CONDUIT RULES (FOR BOTH ROOFTOP AND GROUND-MOUNT INSTALLATIONS)

In addition to standard best practices and installation techniques, running conduit for PV systems involves some special considerations:

- Conduit and fittings must be outdoor-rated (watertight).

- Conduit runs can have a maximum of 360 degrees in bends (such as four 90-degree elbows) between boxes or other pulling points (points of access for pulling wires). This is a standard code rule but is especially important to remember when routing conduit around the edge of the roof, along the soffit, down the wall, and so on. A rule of thumb that pros follow is to make every fourth bend in a run a pulling point, using a conduit body, pulling elbow, or junction box.

- Secure conduit with a clamp within 3 feet of all boxes and every 6 feet elsewhere. This guideline is a little stricter than many standard requirements but is a good idea for getting PV systems approved.

- Rooftop conduit may need expansion couplings to tolerate high temperatures, especially if runs are longer than 10 feet along the roof.

Component Connections

The devices and wiring configuration at each of the remaining components of a PV system are specific to each system installation. Your electrician will make the necessary calculations regarding wire size, overcurrent and lightning protection, grounding, and other essential requirements. Here, we'll look at a basic sample of each component and introduce some of the common elements to help you understand how the final connections complete the system.

NOTE: These are intended as generic examples only and should not be used as a guide for wiring your own system.

DC Combiner Box (String-Inverter Systems)

Combiner boxes allow you to join multiple-module series-strings in parallel to minimize the number of wires that make the trip down to the ground-level components. Combiners with disconnect capability also serve as the disconnect for rapid-shutdown systems. The basic configuration of a combiner box is similar to an electrical subpanel or small breaker box. The home run cables from the series-strings connect to fused breakers and a neutral **bus bar** in the box. The array ground wire connects to a ground lug or grounding bus bar, which also grounds the box. With most home PV systems, the outgoing wires include a positive (current-carrying) wire, a negative wire, and a ground wire, plus wiring from the rapid-shutdown control box.

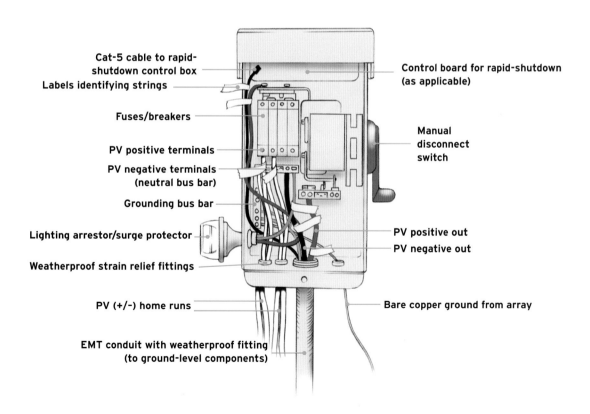

Cat-5 cable to rapid-shutdown control box
Labels identifying strings
Fuses/breakers
PV positive terminals
PV negative terminals (neutral bus bar)
Grounding bus bar
Lighting arrestor/surge protector
Weatherproof strain relief fittings
PV (+/-) home runs
EMT conduit with weatherproof fitting (to ground-level components)

Control board for rapid-shutdown (as applicable)
Manual disconnect switch
PV positive out
PV negative out
Bare copper ground from array

AC Junction Box (Microinverter Systems)

An array junction box on a typical microinverter system receives the AC trunk cables and ground wire from the array. With standard 240-volt, single-phase wiring, the outgoing wires include two positive (current-carrying) wires, each one carrying 120 volts; one neutral wire; and a grounding wire. The ground wires should connect to the metal box to ground it.

NOTE: You may have secured the AC trunk cables to your junction box during the prewiring (as directed by the manufacturer), but from this point forward the connections to the trunk cables and all AC wiring and installations should be done by a licensed electrician.

Rapid-Shutdown Control (String-Inverter Systems)

The rapid-shutdown control box is the hub of the rapid-shutdown system, connecting to both the disconnecting combiner box at the array and the DC-AC inverter. The box includes a manual switch that allows emergency responders to activate the shutdown. In addition to the circuit wiring, the control box is linked to the system components with Cat-5 cable for communications.

REMEMBER: Systems with microinverters do not need separate rapid-shutdown equipment because the microinverters automatically shut off the solar-generated power at the modules.

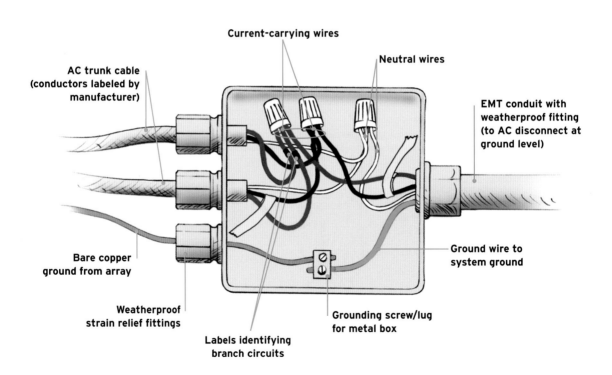

Current-carrying wires

Neutral wires

AC trunk cable (conductors labeled by manufacturer)

EMT conduit with weatherproof fitting (to AC disconnect at ground level)

Bare copper ground from array

Ground wire to system ground

Weatherproof strain relief fittings

Grounding screw/lug for metal box

Labels identifying branch circuits

This demonstration setup shows a Birdhouse rapid-shutdown control box next to a disconnecting combiner box. The blue wires are Cat-5 cables, which transmit the trip signal between the control box and the disconnect.

DC–AC String Inverter

String inverters receive DC electricity from the PV array and output AC electricity to the rest of the system. AC and DC power operate at different voltage and current levels, requiring different wiring for each side of the inverter. Today's string inverters include integrated DC disconnects that are prewired at the factory. If you happen to have an older-style inverter without a DC disconnect, then your system must include a separate DC disconnect unit installed between the array and the inverter. The string inverter itself is usually mounted on a wall outside the house or on an interior wall in the garage or electrical room.

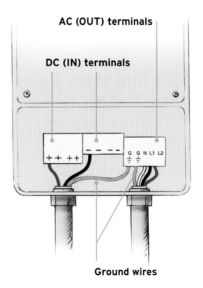

AC (OUT) terminals

DC (IN) terminals

Ground wires

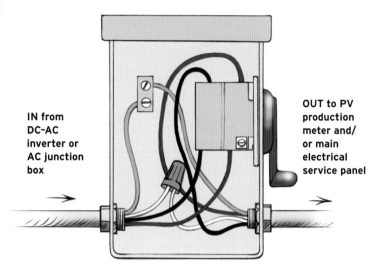

IN from DC-AC inverter or AC junction box

OUT to PV production meter and/ or main electrical service panel

AC Disconnect

The AC disconnect is a standard electrical device that your electrician can get through a local supply house. It connects to the AC wiring from the DC–AC inverter (for string inverter systems) or to the wiring directly from the array AC junction box (for microinverter systems). The disconnect includes an internal breaker for overcurrent protection and a large manual "knife blade" switch on the box exterior that can be operated without removing the box cover. The switch can be locked to secure it in the off position for service work or any other reason.

Meters and Main Electrical Service Panel

The last three components to be connected in a grid-tied system are the PV production meter, the home's main electrical service panel, and the utility net meter. **Production meters** record all solar-generated energy and are used in some areas to determine SRECs (see page 96) and/ or surplus energy for buyback or net metering programs (see page 95). Production meters may be required by your utility and/or state or local

energy program but otherwise may be optional. Where required, the utility or local authority specifies the type of meter to be used (such as a "revenue-grade" meter) and any installation and signage requirements.

Connection to the **main electrical service panel** typically is made with a double-pole (DP) "backfeed" breaker added to an open slot or two on the existing panel. If there isn't enough space for the new breaker, the electrician can install a subpanel next to the main panel to house the PV system breaker. A label indicates that the breaker is for the PV system. The capacity and type of breaker will be determined by your electrician. Some PV systems have more than one breaker.

The utility (or one of its subcontractors) installs its **net meter**, usually replacing the old standard meter, on the grid side of the main electrical service panel and hooks you up to the grid at the same time. This is the final connection in the PV system installation. It usually is done only after the final inspection is passed and all the utility paperwork has been processed.

Standard meters run in one direction only, while net meters run in two directions. If you were to operate your PV system with a standard meter, the solar energy you produce would make your meter run forward faster, as though your house were using more electricity. (Yes, people have tried to circumvent the utility and run their PV systems without a net meter only to find their electricity bill going up rather than down.)

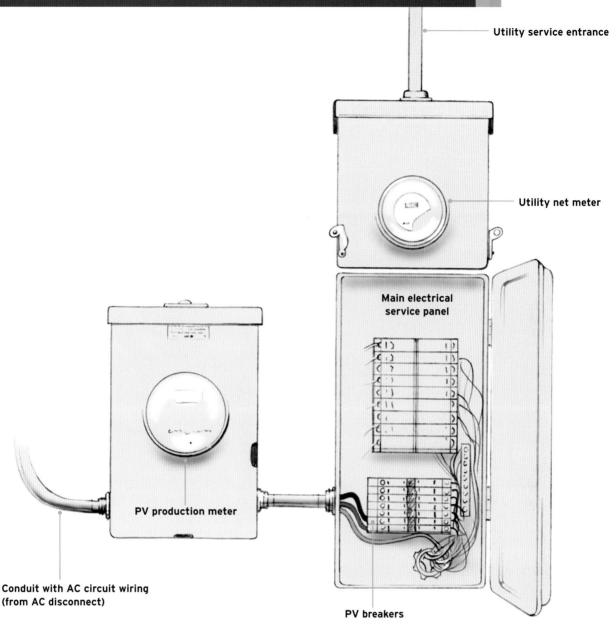

Utility service entrance

Utility net meter

Main electrical service panel

PV production meter

Conduit with AC circuit wiring
(from AC disconnect)

PV breakers

WARNING: The service lugs in the main electrical service panel remain live **– carrying deadly voltage –** even when the panel's main circuit breaker is shut off. The lugs are covered by the box's *dead front cover*, the metal panel on the interior of the box, accessible only when the box's door is open. The dead front cover conceals the internal wiring, as well as the lugs, and has a rectangular cutout that allows access to the breaker switches. **Do not remove the dead front cover.** This box should be worked on only by a licensed electrician.

SIGNAGE

Requirements for safety signage are established by the local building department and can be very specific. It's not unusual to have custom stickers or placards made to satisfy the inspector. Warning signage must have either black lettering on a yellow background or white lettering on a red background. Common messages on warning labels include the following:

- Photovoltaic System Connected (Active)

- Solar-Electric (PV) System

- This system is Fed by a Solar Array as well as the Utility

- Dual Power Supply/Sources: Utility Grid and PV Solar Electric System

- Photovoltaic System AC Disconnect

Final Inspection and Turning On Your PV System

When your electrician is done connecting your system to the main electrical service panel, you will call for a final inspection from the building department. Double-check your permit packet to make sure everything is in place for the final inspection, including warning signage. If the inspector needs to examine any rooftop components, set up a ladder, following all essential ladder safety guidelines (see page 100).

After you've passed the final inspection, notify your utility provider. They will send out one of their electricians or subcontractors to install a net meter, which connects your PV system to the utility grid via the main electrical service panel. When that work is done, the utility sends you a Permission to Operate (PTO) letter, by mail or email, granting you official approval to turn on your system. Be sure to take that official approval seriously; **do not** turn on your PV system until you have the PTO notice in hand.

For the last step, it's all you, baby: turn on your system following the instructions on pages 173–175 for your system type.

Congratulations! You're a solar photovoltaic clean-energy producer!

9 Off-Grid System Design

GOAL

Create an entire off-grid PV system on paper

IF YOU'RE SERIOUS ABOUT OFF-GRID POWER, you won't be surprised to hear that designing for off-grid PV systems is a little more complicated than for grid-tied systems, but it's probably simpler than you think. The two system types have a lot in common, and, to counteract the three main things that are added on the off-grid side — batteries, charge controller, generator — there are two things you don't have to worry about: meters and hassles with the utility (not to mention utility bills). With the similarities in mind, you'll need to have a thorough understanding of the design fundamentals and processes discussed in chapter 4 as a foundation for learning off-grid design and how it differs from grid-tied design. So if you just skimmed that chapter (or skipped it altogether) on your way here, we're afraid you'll have to go back and digest chapter 4 first. If you don't, this discussion of off-grid systems is sure to give you a stomachache.

NOTE: *When it comes to the installation of the ground-level electrical components (everything but the array), we offer the same recommendation that we did for the grid-tied folks: hire an electrician or professional solar installer to make all of the final wiring connections. That includes the batteries, which in some ways are the most dangerous parts of all PV installations.*

Off-Grid Basics

To appreciate the primary difference between off-grid and grid-tied PV systems, imagine you're planning to take everyone in your household on a long camping trip. (Hear us out.) One option is to go car-camping, where you park your vehicle right at the campsite. You'll spend your days and nights outdoors, but if you run out of food, beer, toilet paper, or other essentials, you'll have plenty of backup supplies packed in the car, just a few steps away.

The other option is a wilderness trip where you carry everything you need to survive deep into the woods in a backpack. If you plan carefully and make sure nobody eats or drinks too much, you'll do just fine. But if supplies run low, you'll have to cut back on your consumption. And if the supplies run out, they run out. You're camped too far from the car to hike back, and you didn't fill it with extra supplies anyway.

You get the idea. Planning an off-grid PV system, like wilderness camping, requires more care and thoughtful examination of what you need on a daily basis. The only thing missing from the comparison, in the off-grid scenario, is a generator for replenishing the energy supply if things get dire. For the sake of the camping metaphor, we'll call it a fishing pole.

It's a simple matter of supply and demand. Sizing an off-grid system starts with a careful examination of your household electrical demand — your daily electrical loads. You'll learn how to calculate loads on page 152. In terms of equipment, let's take another look at the basic elements of an off-grid PV system, since there are additional components beyond those necessary for a grid-tied system.

PV ARRAY: The modules and module support structure are the same for off-grid and grid-tied systems, but with an off-grid design, the module series-strings tend to be much smaller, typically with strings of three or four modules each, as we'll discuss in Array Layout (page 154).

RAPID SHUTDOWN: Off-grid systems use string inverters (not microinverters), which means they need a dedicated system to meet rapid-shutdown requirements. This usually takes the form of a rapid-shutdown control switch that communicates with a disconnecting combiner box at the array, as with a standard grid-tied string-inverter system.

CHARGE CONTROLLER AND BATTERIES: In an off-grid system, the solar-generated DC power from the array passes through a charge controller on its way to the batteries. The **charge controller** regulates the flow of power and steps down the voltage to the proper level for the battery bank design (see Battery-Bank Wiring, page 160). The charge controller is essential for safe charging and for battery health. It turns on when the battery voltage is low and the battery bank needs charging. It turns off when the battery voltage is high and the batteries are fully charged.

DC–AC INVERTER: The stored DC power in the batteries runs through an inverter to be converted to AC power for use in the household electrical system. Remember that inverters for off-grid PV systems are called **stand-alone inverters**, while those for grid-tied systems are called "grid-tie inverters." Conventionally, microinverters are not used for off-grid systems

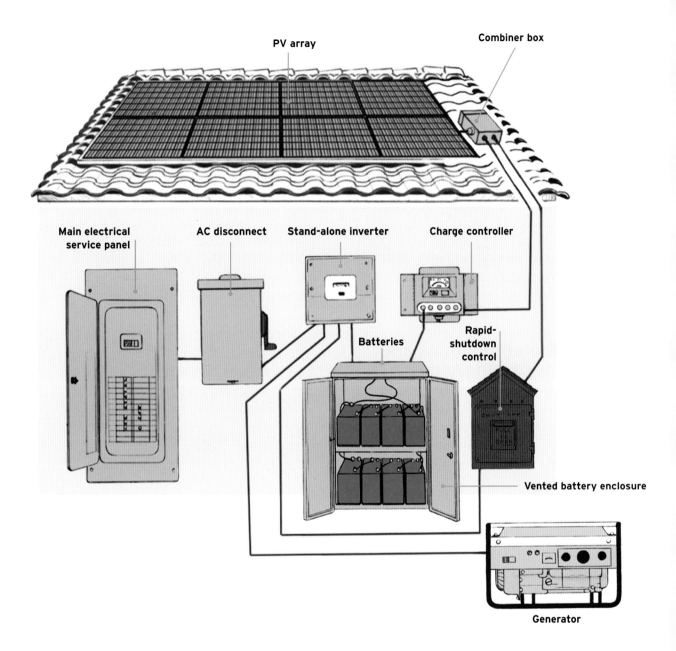

PV array

Combiner box

Main electrical service panel

AC disconnect

Stand-alone inverter

Charge controller

Batteries

Rapid-shutdown control

Vented battery enclosure

Generator

Off-grid PV systems have the most in common with grid-tied systems with string inverters. The main differences in equipment are the inclusion of a charge controller, battery bank, a generator (typically), and a stand-alone inverter (in place of a grid-tied inverter). Off-grid systems also stop at the home's electrical service panel, since there's no utility meter or grid connection.

OFF-GRID BASICS

151

because the batteries require DC power, and microinverters convert to AC at the module. However, there are *AC-coupled* systems that use them (in a setup sometimes called a "micro-grid"), but this entails unnecessary complication for most single-home systems.

AC DISCONNECT: On the AC side of things, off-grid systems have only an AC disconnect and the home's main electrical service panel, both of which are the same as those used with grid-tied systems. Of course, no production meter or utility meter is required, since you are not connected to the utility grid.

GENERATOR: Off-grid households that rely on solar power almost always have a generator for backup power when battery storage levels are low and/or the sun isn't shining.

Calculating Loads and Days of Autonomy

Your electrical loads are the cumulative totals of all daily electrical usage in your house. Calculate your loads by multiplying the wattage of each electric appliance and device by the average amount of time it is used in a day. For appliances that are used less regularly, such as a clothes washer, calculate the daily average based on your weekly use. For example, if you run the washer 3.5 hours per week, use a daily figure of 0.5 hour (3.5 ÷ 7 = 0.5). Add

up the usage of all appliances to find the daily total kilowatt-hours (kWh) used in the home. This number is your minimum daily goal for AC power produced by your PV system. During the design you will use PVWatts to determine a DC system size for reaching this goal.

The chart on the following page lists some common household appliances and devices and their approximate wattages and daily usage times (annual averages). This gives you just a snapshot of a household with fairly efficient appliances. As you probably know, there are manufacturers that specialize in ultra-efficient appliances for off-grid homes. Investing in just one ultra-efficient major appliance, such as a refrigerator or stove, can make a huge difference in a household's daily load.

Determine the wattage of your own appliances by looking on the manufacturer's nameplate. If it doesn't list the wattage but gives you the amps, multiply by 120 (for standard-voltage appliances, or by 240 for dryers, electric stoves, and other high-voltage appliances). For example, if an electric fan is rated for 3.0 amps, its wattage is 3 amps × 120 volts = 360 watts.

Wattage is the measure of instantaneous power consumption of an appliance – in the same way that it's the measure of power output from a PV module at a given point in time. To convert

TIP

LOWEST SUN, HIGHEST LOADS

The short supply of sun and high electricity demand (heating, lighting) in winter has a double-whammy effect on a PV system's supply and a home's demand. It's best to take a conservative approach and design your off-grid system for the time of year when the sun hours are at their lowest (December) and your electrical loads are at their highest. You should also look at cooling loads, if necessary, in the warm summer months (if your home has air conditioning, for example).

ELECTRICAL LOADS (SAMPLE CALCULATION)

LOAD (APPLIANCE)	WATTAGE	# OF UNITS	DAILY USAGE (HRS./DAY; ANNUAL AVERAGE)	TOTAL DAILY LOAD (KWH/DAY)
Refrigerator	250	1	24	6.0
Stove/Range	1,000	1	0.7	0.7
Microwave oven	1,000	1	0.2	0.2
Well pump	700	1	0.75	0.53
Furnace blower fan	350	1	2	0.7
Woodstove circulation fan	200	1	2	0.4
Ceiling fan	120	1	2	0.24
TV/Stereo	200	2	5	2
Lighting (LED or CFL)	12	5	5	0.3
Computer	150	1	5	0.75
Chargers	25	2	5	0.25
Clothes washer/dryer combo	450	1	0.5	0.23
			Total Daily Load	**12.29 kWh/day**

wattage to actual energy usage, you have to add a factor of time: a 360-watt fan running for 3 hours uses 1,080 watt-hours. Divide the watt-hours by 1,000 to find the kilowatt-hour (kWh) usage: 1,080 ÷ 1,000 = 1.08 kWh. Use the kWh values for totaling your household electrical loads.

Days of autonomy (DOA) is a critical design factor for off-grid systems. It answers the question "How many days can I go with no sun?" In other words, if it's the dead of winter and the forecast shows nothing but snow or dense cloud cover for the near future, how long do you want your batteries to meet your daily loads before you run out of power?

Days of autonomy are determined by the capacity of your battery bank. You can choose the number of days you like, but the standard target is three days, assuming the system includes an electrical generator for backup power during extended periods of cloudy weather. A three-day target seems to strike the right balance between cost and benefit for most off-grid homeowners. There is no rule against designing for more than three days; it just requires more money for a larger battery bank. You will apply your target number of days of autonomy when sizing your battery bank.

Array Layout

Back in chapter 4, we explained that design for the array (including the modules and module support structure) is the same for grid-tied and off-grid systems. To clarify: The **physical** layout of the modules and all the support structure calculations can be exactly the same, but the **electrical** layout is different.

With a grid-tied string-inverter system, the length of the series-strings is ultimately limited by the maximum allowed input DC open-circuit

voltage (V_{oc}) rating of the string inverter, and 8 to 12 modules in a series-string is common. With off-grid systems, the series-strings are limited by the charge controller. Most charge controllers can handle strings of no more than three or four modules in series. For example, if you have 12 modules total, you typically would wire the array in four parallel strings of 3 modules each. Many newer charge controllers can handle 4 modules in series (three parallel strings of 4 modules each for a 12-module array), and some can do several more in series, with ratings up to 600 volts maximum V_{oc}.

Choose the charge controller rating that best fits your array size. Remember that the physical layout of the modules doesn't have to match the electrical layout. For example, your 12-module array could be laid out in two physical rows of 6 modules each and still be wired as four strings of 3 modules each.

Charge Controllers

Charge controllers are needed for off-grid PV systems because the module strings put out much higher voltage than the batteries can handle. In the sample design shown later in this chapter, the module series-strings have a rated output of about 91 volts, while the battery bank is rated at 48 volts (see Battery-Bank Wiring, page 160). The charge controller receives 91 volts (under ideal conditions) from the array and delivers 48 volts to the batteries. In addition, it regulates the amount and rate of charging that the batteries require at all times.

Charge controllers come in many types, with different ranges of cost and capability. The only type that's highly recommended for an off-grid PV system is an **MPPT charge controller**, which can accept an input voltage (from the array) that's higher than its output voltage (the voltage that goes to the batteries), as discussed previously. The other type is a **fixed-voltage charge controller**, which requires that the array voltage be the same as the battery voltage. This type is used only with very small PV systems and usually is not suitable for home power systems.

Maximum power point tracking (MPPT) capability on a charge controller is similar to MPPT with inverters (and DC optimizers), discussed in chapter 3. In this case, the charge controller (not the inverter) constantly monitors the voltage and current of the modules and adjusts the levels as needed to maintain the maximum power point (MPP). Because the most dramatic adjustments are needed during less-than-ideal sun conditions, extreme temperature conditions, and low battery conditions, MPPT charge controllers really earn their keep in low light (cloudy weather), when the modules are partially shaded, or during any extremes in solar irradiance or temperature.

Today's MPPT charge controllers have computer-interface capability, allowing you to monitor your system remotely from a computer or other device.

Batteries

The subject of batteries for PV systems could easily fill a book on its own. There are many types, sizes, prices, characteristics, and behaviors of batteries. There are also countless personal experiences and opinions out there, not to mention the fact that the same batteries often perform differently for different PV systems, and under varying solar irradiance, temperature, and electrical load conditions. With that in mind, we'll discuss the basics to get you started. And the reality is, batteries last only about 5 to 8 years on average (although some newer types are predicted to last 10 to 15 years), so you'll likely go through several sets during the life of your PV system, learning what works best along the way.

One universal rule is to use only deep-cycle batteries, which are designed for the regular deep discharges of a PV system. By contrast, starter batteries, such as car batteries, are designed for quick bursts of discharge to start the engine, followed by immediate recharging while the engine is running. The deep discharge of a PV charge-recharge cycle would quickly kill a car battery.

TYPE. There are several options for suitable battery types, outlined in the chart on the next page. The most commonly used type is lead-acid, thanks to its attractive compromise of price and longevity. Flooded lead-acid (FLA) batteries tend to offer the most storage capacity for the money and are the workhorse of the off-grid PV system, but they require regular maintenance to ensure longevity. Sealed lead-acid batteries cost a little more per storage capacity but require very little maintenance (and are often called "maintenance-free" batteries). The two common types of sealed batteries are absorbed glass mat (AGM) and gel-cell batteries. Two other types of deep-cycle batteries, nickel-iron (NiFe) and lithium-iron phosphate, (LiFeP) – offer considerably greater cycle life and therefore longer lifetimes and may represent the future of off-grid storage systems.

MAINTENANCE. All batteries should be monitored regularly to make sure they are charging and discharging properly, but in terms of manual maintenance this usually means flooded lead-acid batteries. Maintenance tasks are simple and safe (if done properly) but must be performed on a regular basis

(see Battery Maintenance for Off-Grid Systems, page 178.)

COST. Battery banks can cost anywhere from around $2,000 to upward of $10,000. Some of the priciest batteries are designed to last 15 years or more, but many types have the standard life expectancy of 5 to 8 years. In any case, battery life is never guaranteed and can be affected by factors such as temperature (some batteries have a shorter life if they're subjected to high temperatures) and **depth of discharge**

BATTERY COMPARISON CHART*
*Values based on specific battery models; will vary by manufacturer, model, and operating conditions.

TYPE	VOLTAGE (TYPICAL)	CAPACITY (20-HOUR RATING; IN AMP-HOURS)	CYCLE LIFE — # OF CYCLES (@50% DEPTH-OF-DISCHARGE— DOD)
Flooded lead-acid (L-16)	6-volt; also available in 2-volt	370	1,650
Flooded lead-acid ("golf cart")	6-volt; also available in 8-volt	225	1,650
Sealed AGM (lead-acid)	12-volt; also available in 2- and 6-volt	210	1,350
Gel-cell (lead-acid)	6-volt; also available in 12-volt	210	1,000
Nickel-iron (NiFe) "Edison battery"	Assembled in 12-, 24-, and 48-volt packages	335	11,000 (@ 80% DOD)
Lithium iron phosphate (LiFeP)	Assembled in 12-, 24-, and 48-volt packages	300	3,000

(most deep-cycle batteries live longest when discharged no more than 50%). This is why it's often recommended that off-grid beginners choose a "starter set" of relatively low-cost batteries with a decent service life, such as a flooded lead-acid or a maintenance-free sealed battery. If they make some mistakes and the batteries don't last very long, it's better to have spent $3,000 than $10,000.

PHYSICAL SIZE. Battery size, shape, and weight usually aren't important criteria for home power systems, since the batteries permanently reside in some kind of utility room, garage, or shed and need to be moved only when they get replaced. Be aware that batteries are heavy (ranging from about 60 pounds each for small versions to 400 to 600 pounds for industrial batteries) and that all batteries must be housed in a lidded, ventilated enclosure (see Battery Enclosures, page 158). The difficulty of getting heavy batteries in and out of an enclosure is something else to keep in mind so accessibility is important as well.

SIZE (APPROX.; LENGTH × WIDTH × HEIGHT) & WEIGHT	PRICE	REGULAR MAINTENANCE?	NOTES
11¾ × 7 × 17½ inches 118 pounds	$350–$400	Yes	Larger of the standard flooded lead-acid batteries used for PV; for many years considered the industry standard for off-grid storage
10⅜ × 7⅛ × 11¾ inches 60–70 pounds	$175–$200	Yes	Smaller, "golf-cart" size of flooded lead-acid battery; models designed for PV have relatively long discharge rates for longer overall battery life
21 × 8¼ × 8½ inches 133 pounds	$575–$650	No	Standard option for maintenance-free lead-acid battery
9⅝ × 7½ × 11⅝ inches 69 pounds	$340–$380	No	Alternative to sealed AGM; requires slow charging rate but usually this is not a problem with PV
55 × 11 × 17¾ inches 480 pounds	$3,000	Yes	12-volt battery made up of 10 battery cells, each 1.2 volts; manufacturers claim no loss of life with 80% DOD compared to 50% DOD
14⅛ × 8½ × 12¾ inches 86 pounds	$2,500	No	Considered by some to be the future of PV batteries; not the same as lithium-ion batteries

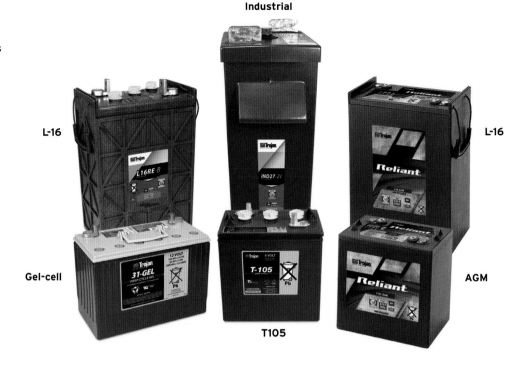

Various Types of Lead-Acid Batteries

Industrial

L-16

L-16

Gel-cell

AGM

T105

CAPACITY. Battery capacity is measured in amp-hours. Standard deep-cycle batteries for PV systems might have an individual capacity ranging anywhere from 100 to 500 amp-hours each. To see how amp-hours translate to usable watt-hours or kilowatt-hours, multiply the amp-hour rating by the battery voltage. A 12-volt, 200-amp-hour battery can discharge 2,400 watt-hours, or 2.4 kWh, of energy – minus efficiency losses from converting the battery's DC power to AC power (based on the inverter efficiency) as well as wire losses and battery discharge losses.

CYCLE LIFE AND DEPTH OF DISCHARGE. Cycle life is an estimate of how many times a battery can complete single cycles of discharging and recharging. Under normal conditions, batteries in off-grid systems complete one cycle per day. Cycle-life ratings for deep-cycle batteries can range from about 500 to about 3,000 cycles. Deep-cycle batteries last longer if they are discharged no more than 50% to 80% with each cycle, depending on the battery type. Lead-acid batteries should be discharged to 50%, while nickel-iron (NiFe) batteries may tolerate up to 80% discharge.

No batteries should be discharged more than 80%, which significantly shortens their life. Note the manufacturer's recommended depth of discharge (DOD) when comparing cycle life and battery capacity.

Battery Enclosures

Battery banks must be kept in solid enclosures for two very good reasons: (1) batteries contain **lethal levels of electricity** that can be unleashed simply by touching two battery terminals with any conductive material (a metal tool, utensil, toy, etc.), and (2) batteries emit **explosive hydrogen gas** during normal charging and discharging. For these same reasons, enclosures should have a lockable lid, and they must be ventilated to exhaust the hydrogen gas to the outdoors.

Flooded lead-acid batteries require *active* ventilation provided by an electric fan that pulls air through the ventilation pipe from the battery enclosure to the outdoors. Sealed batteries also need ventilation, but typically this can be *passive*, provided by a ventilation pipe without a fan. All ventilation pipes must terminate outdoors.

PVC pipe vent – 2" diameter or as required; include in-line fan for FLA batteries

Caulk around pipe to seal

¾" plywood box and lid

Lid hinged here

Lid can be flat (level) or angled down

Caulk to seal

Locking hasp and padlock

Box height – height of batteries plus 6"

EMT conduit for battery cables

1¼"-diameter vent hole – 3 total – for makeup air

Box depth – add about 2" at front and back for room to move batteries; make sure you can reach all batteries for maintenance

Box length – include a few extra inches of space at both ends

Paint box interior and exterior

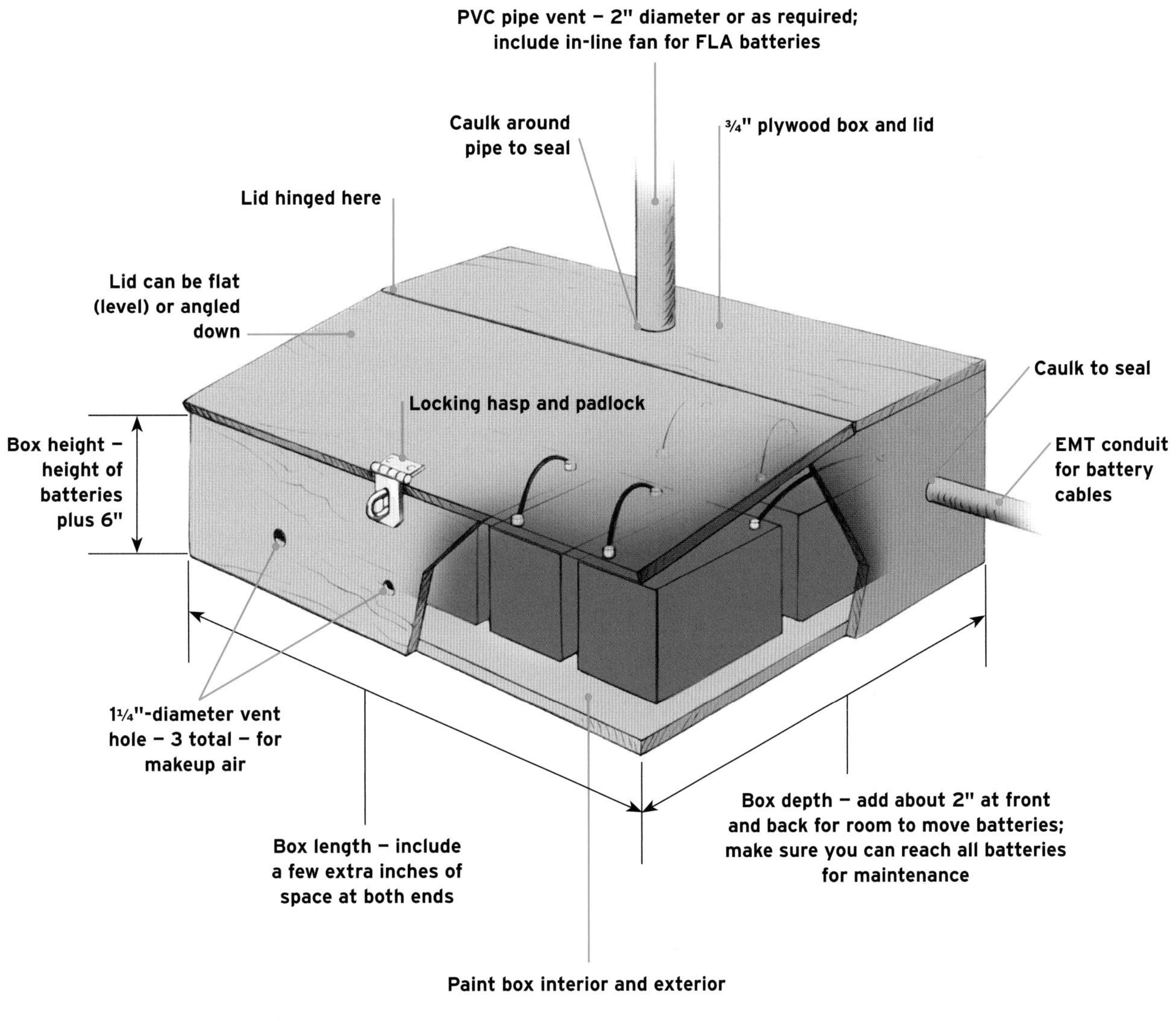

WARNING: Don't make the mistake of omitting an enclosure on the reasoning that only adults live in the house and/or everyone knows about the dangers of batteries. That's like building a high deck without a railing and telling everyone to watch their step. It also assumes that children (or pets) will never, ever visit your house. Besides, battery enclosures are required by the National Electrical Code.

BATTERIES

Batteries work best and last longest when kept at a temperature between 50 and 75°F. This is a third reason to keep them in an enclosure. The box shields the batteries from direct sunlight, and the enclosed space helps keep all of the batteries at the same temperature. Insulating an enclosure is recommended, but be aware that insulation merely slows the transfer of heat (or cold) into and out of the box; it does not raise or lower the temperature in the box. One way to combat cold in a frigid location is to install a small incandescent light-bulb – no more than 40 watts – controlled by a temperature-sensor switch inside the enclosure so the bulb turns on only when it's needed.

It's best if the enclosure is in a room that stays within 50 to 75°F. However, if the floor gets very cold, it's a good idea to build the box atop a base of 2×4s insulated with rigid foam insulation to provide a thermal break between the box and the floor.

Enclosures for residential systems often are simple plywood boxes custom-sized for the battery bank. Suggestions for basic box construction are shown in the battery enclosure construction drawing on page 159. To this you might add rigid foam insulation, if applicable (some experts also like to include a layer of noncombustible cementboard), along the interior walls. Follow the requirements of the local building department when constructing your own enclosure.

Another type of DIY enclosure includes a framework made of metal framing channel (Unistrut is a common brand) with plywood shelves and front-opening doors or removable Plexiglas sheets.

Alternatively, some battery manufacturers offer metal boxes you can use for enclosures.

Depending on the battery size, these may hold about four to eight batteries and may be stackable, with a lid that opens at the front.

Battery-Bank Wiring

Battery banks are typically wired for 12, 24, or 48 volts, a figure known as the *nominal voltage*. Remember that when wiring modules or batteries in series, the voltages add up. A series-string of eight 6-volt batteries (or four 12-volt batteries) has a nominal voltage of 48 volts. This is the most commonly used nominal voltage for residential off-grid PV systems.

Battery banks with multiple series-strings are wired similarly to multiple module strings. The batteries in each series-string are connected positive (+) to negative (−). At one end of the strings, the negatives of the two end batteries are joined. At the other end, the positives of the two end batteries are joined. This connects the two series-strings in parallel, so the amp-hour ratings add together. The positive and negative from the ends of the strings connect to the DC–AC inverter.

Battery banks can be configured with one to four series-strings of batteries wired in parallel. It's not a good idea to combine more than four series-strings in parallel because this can lead to uneven charging, which shortens battery life.

The wiring diagram opposite shows a battery bank with 16 batteries wired in two series-strings of 8 batteries each. The individual battery voltage is 6 volts, and each has an amp-hour rating of 380 amp-hours. Each series-string therefore has a voltage of 48 volts and an amp-hour capacity of 380 amp-hours. When the strings are wired together in parallel, the amp-hour capacity doubles, for a total of 760 amp-hours at 48 volts. To convert this to kilowatt-hours, multiply the amp-hour value times the voltage, and then divide by 1,000: 760 × 48 ÷ 1,000 = 36.48 kWh of total battery storage capacity.

Now, before you get excited and start thinking that a battery bank of this size will give you over 36 kWh of power each day, remember that deep-cycle batteries (and pretty much all batteries) should not be fully discharged. So you must factor the recommended maximum depth of discharge (DOD) into your calculations. For lead-acid batteries, the recommended DOD is usually 50%. Therefore, you have to double the required battery capacity to meet your power goals because you're discharging the batteries no more than 50% before recharging them. The usable storage capacity of the battery bank above is 36.48 ÷ 2 = 18.24 kWh. We will perform all of these calculations in our sample design on pages 164-169.

WARNING: Lethal currents can be discharged when wiring battery banks. It is **highly recommended** that this be done by a licensed electrician or solar professional with battery experience.

Stand-Alone Inverters

Inverters for off-grid PV systems are similar to grid-tied string inverters but are designed for much lower DC input voltage — typically 12, 24, or 48 volts. This is because they receive their input DC voltage from the battery bank, not the PV array. Manufacturers use a standard coding for model names that indicates the maximum power rating and the nominal battery voltage. There are four numbers in the code: the first two indicate the power rating, and the last two indicate the battery voltage. For example, a model with the code 3648 is rated for 3,600 watts, or 3.6 kW, at a nominal battery-bank voltage of 48 volts. A model with the code 2812 is suitable for a 2.8 kW, 12-volt system.

Some off-grid systems require more than one inverter, and some manufacturers of stand-alone inverters offer a wall-mounting plate for 1, 2, 3, or 4 inverters as well as multiple charge controllers.

The AC output of a stand-alone inverter can be 120-volt only, or it can be 120/240-volt, known as **split-phase**, which allows the system to power 240-volt appliances or high-demand

Battery-Bank Wiring

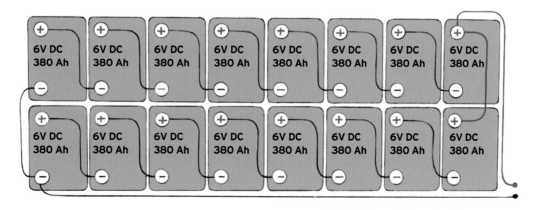

Total Voltage: 48 (nominal)
Total Amp-Hours: 760 (nominal)

devices like well pumps. Like standard utility-powered household electrical systems, split-phase inverters use the combination of two 120-volt circuits (sometimes called "legs") to provide 240-volt power when needed. Loads that require 120-volt power can be supplied by either of the 120-volt legs. However, due to how these inverters are designed, it's important to balance the house's 120-volt loads between the two legs, because each leg typically has an actual power output that's about 25% less than its rated output. If you're planning to use a 120/240-volt inverter, be sure to check the product specs and discuss your application with the inverter manufacturer.

Off-grid PV systems that include backup generators usually have stand-alone inverters with integrated chargers. These allow you to charge the batteries from the generator without having to use a separate charging unit. With some off-grid configurations, the generator (instead of the batteries) is used to power large loads in addition to recharging the batteries. This often requires a relatively large generator, particularly if it must cover large surge loads, such as starting up a well pump. An alternative is to use a smaller generator that is sized to handle normal household loads while it's charging the batteries. If there's a demand for a surge load, the inverter draws a temporary boost of power from the batteries to assist the generator (the batteries have some reserve power even while they're getting charged). This configuration, sometimes called "generator support," requires an inverter with an output that can cover the household's largest loads.

Generators

Most off-grid PV systems include a generator for backup power when the PV system can't meet the demands of the household electrical loads, usually during the low-light days of winter or extended cloudy periods. Generators also are useful for equalizing batteries (see Battery Maintenance for Off-Grid Systems, page 178) and for powering (or helping to power) significant loads that would otherwise use a lot of stored power in the batteries.

There are many factors to consider when choosing a generator, including the size of the battery bank, the household loads, and the AC output of the PV inverter. If the house has only 120-volt loads, you can use a generator that outputs at only 120 volts. But if the house has some 240-volt loads, the generator must have both 120- and 240-volt output. In this case, it's important to check the generator's output ratings carefully to understand the actual output at both 120 and 240 volts. As with stand-alone 120/240-volt inverters (page 161), there may be a reduction in output when drawing 120-volt power. Generators typically are sized to cover a little more power than the PV inverter's full charging capacity plus any household loads that may run concurrently with battery charging.

Fuel types for generators include propane, natural gas, gasoline, and diesel fuel. The most commonly used type for off-grid systems is propane, for a few reasons. Off-grid homes often use propane for cooking and heating, so the main supply and storage equipment is already in place. Propane — also called liquefied petroleum (LP) gas — can be stored for long periods and can operate at low temperatures. By contrast, gasoline is more difficult and hazardous to store, and it has a relatively short shelf life. Diesel is used for many quality, high-powered (and pricey) generators, but diesel fuel is problematic in sub-freezing temperatures, and it doesn't burn as cleanly as propane or natural gas.

Generators are available with remote and/or auto-start features. Remote capability allows you to monitor system functions from inside your house and to start and stop the generator remotely. (They are usually installed in an outbuilding to remove the risk of deadly exhaust in the house and to mitigate noise and odors.) An auto-start (auto generator start, or AGS) function allows a generator to be started automatically by the PV system inverter whenever the battery-bank charge level dips below a set point. Keep in mind that auto-start means the generator may run when you're asleep or away from home, when you're not able to keep an eye on the system in case something goes wrong. They can also fail to start due to various problems, such as a dead starting battery, out of gas, bad gas, and so on.

With the standard configuration, a backup generator is connected to the PV system's stand-alone inverter (with integrated charger). The inverter/charger has an internal transfer switch that controls the flow of AC power from the generator to the house's main electrical service panel. It also converts the AC power from the generator to DC power for charging the battery bank. The generator connects only to the inverter/charger and not to the main electrical service panel.

WARNING: Remember, all AC wiring should be installed by a licensed electrician. This includes batteries, generators, and related wiring.

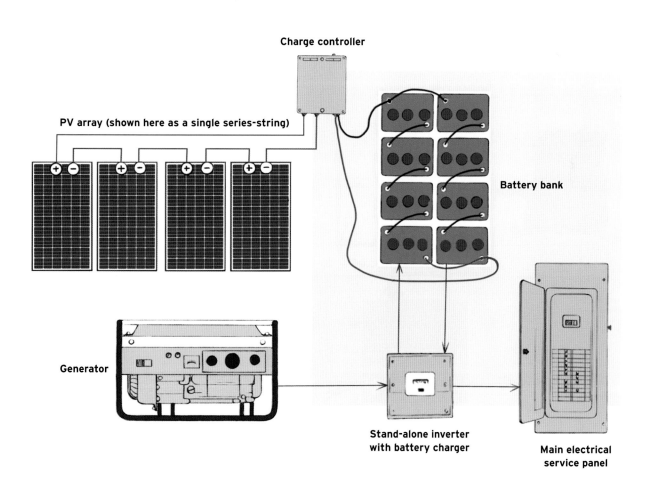

Charge controller

PV array (shown here as a single series-string)

Battery bank

Generator

Stand-alone inverter with battery charger

Main electrical service panel

Off-Grid System Design

As in chapter 4, here we'll walk through the steps for creating the array layout and mechanical and electrical specs for a sample off-grid system. We'll skip over much of the mechanical design for the array because these steps are covered in chapter 4 and are the same for grid-tied and off-grid systems. Also the same are the electrical components for rapid shutdown control, the AC disconnect, and the connection to the home's main electrical service panel.

Most of what's unique to off-grid PV design is in the electrical calculations for sizing the system and choosing the equipment. The basic process involves five main steps:

1. LOADS: calculating household electrical loads

2. DC POWER GOAL: sizing the PV array using PVWatts

3. PV MODULES: running module specs and determining the array's physical layout

4. DC SYSTEM SPECIFICATIONS: electrical calculations for module series- and parallel-strings, charge controller, and stand-alone inverter

5. BATTERIES: sizing the battery bank using battery model specs

The sample design shown here is a 48-volt, 3 kW system with a rooftop array of 12 modules wired in four series-strings of three modules each. The battery bank has 32 flooded lead-acid batteries (L-16 type) and is capable of providing three days of autonomy. It has one DC–AC inverter and one charge controller. As with the sample designs in chapter 4, the sample values are shown in blue type.

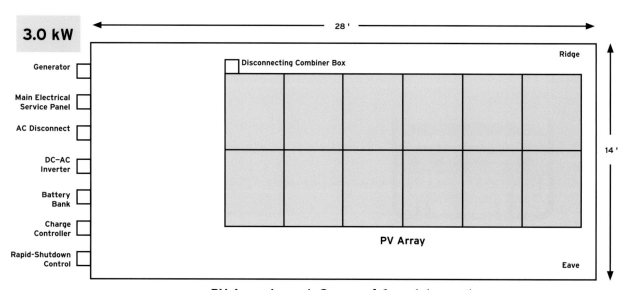

PV Array Layout: 2 rows of 6 modules each

SYSTEM ELECTRICAL SPECS

Module Orientation: Portrait   Azimuth: 205°   Tilt: 23°

12	Total # of PV Modules
3	# of PV Modules in Series
30.3 Volts	PV Module Operating Voltage
90.9 Volts	String Operating Voltage
140.3	Maximum String V_{oc}
4	# of PV Module Parallel Strings
32.9 Amps	Total Current Into Charge Controller
1	# of SA Inverters (Outback VFX3648)
1	# of Charge Controllers (Outback MX60)
90.9 Volts	Charge Controller Voltage IN (from PV array)
48 Volts	Charge Controller Voltage OUT (to Batteries)
6	Individual Battery Voltage
8	# of Batteries in Series
32	Total # of Batteries

PV MODULE SPECS

Model: Helios 6T-250

P_{max} =	250 Watts
V_{pmax} =	30.3 Volts
V_{oc} =	37.4 Volts
I_{pmax} =	8.22 Amps
I_{sc} =	8.72 Amps
Length =	66.1 Inches
Width =	39.0 Inches

24	# of Footers
4	# of Rails
238	Rail Length (inches)
8	# of End Clamps
20	# of Mid Clamps
12	# of WEEBs

BATTERIES

L-16 Flooded Lead-Acid	Type
6 Volts	Individual Battery Voltage
390 Amp-hours	Individual Battery Capacity
4	# of Parallel Battery Strings
1,560 (390 × 4) Amp-hours	Battery Storage
48 Volts	System DC Voltage
74.88 (1,560 × 48 ÷ 1,000) kWh	Total Storage
37.44 kWh	Total Usable Storage (@ 50% Depth of Discharge)
Days of Autonomy:	37.44 (usable storage) ÷ 12.29 (average daily usage) = 3.0 @ 50% DOD

TOTAL ELECTRICAL LOAD

Daily Electrical Load Total: 12.29 kWh

Days of Autonomy Goal: 3.0

TOTAL SYSTEM OUTPUT (PVWatts)

3,000 Watts (DC, peak sunlight)

4,891 kWh (AC, annual)

1. ELECTRICAL LOADS

Calculate your daily electrical loads for your entire household as described in Calculating Loads and Days of Autonomy (see page 152). Double-check (and triple-check) your list to make sure you haven't missed any electricity users in the tally. Also remember to consider the time of year with the highest loads. Calculate your total daily load; then multiply the result by 365 to determine the annual load.

Total daily load: **12.29** kWh

Annual load: **4,486** kWh

2. PV ARRAY SIZE

Determine the size of the array (DC power goal) using PVWatts (see page 26). Enter the basic inputs on the PVWatts "System Info" page, using 18% (recommended) for the system losses. Divide the result (the kWh/year value on the Results page) by 365 to find the daily power production; this value must be higher than your total daily electrical load (usage).

In the sample design, a 3 kW (DC) array will produce 4,891 kWh/year, according to PVWatts, for this selected location, array azimuth, and tilt.

4,891 ÷ 365 = **13.4** kWh/day (this is larger than the 12.29 kWh/day usage in the sample system)

3. PV MODULES

The chosen module is the same model used in the chapter 4 designs, the Helios 6T-250, with the following specs:

P_{max}: **250** watts

V_{pmax}: **30.3** volts

V_{oc}: **37.4** volts

I_{pmax}: **8.22** amps

I_{sc}: **8.72** amps

Length: **66.1** inches

Width: **39** inches

Divide the DC system size (3 kW) by the module wattage to find the number of modules in the array.

3,000 watts ÷ **250** watts per module = **12** modules

NOTE: The total number of modules must be divisible by either 3 or 4 – the maximum number of modules in each series-string determined by the charge controller (charge controllers with higher open-circuit voltage maximum ratings can handle series-strings with more modules). Also, remember that all series-strings must have the same number of modules.

Plan the physical layout of the modules based on the available installation area (step 4 on page 69).

4. DC SYSTEM SPECIFICATIONS

Complete the electrical calculations for the modules, charge controller, and inverter. In this design, the charge controller is rated for series-strings of three modules each.

# of modules in each series-string: **3**

Module operating volts (V_{pmax}): **30.3** volts

String operating voltage: V_{pmax} × # of modules per string

30.3 × **3** = **90.9** volts

Maximum string voltage: Module V_{oc} × # of modules per string × 1.25 (extreme temperature limit from the National Electrical Code)

37.4 × **3** × **1.25** = **140.3** volts

NOTE: This value must not exceed the maximum allowed extreme voltage of the chosen charge controller (150 volts in this example)

# of module parallel strings: **4** (12 total modules, with 3 modules in each series-string, wired in parallel)

Total current into charge controller: I_{pmax} × # of parallel strings

8.22 × **4** = **32.9** amps

NOTE: This value must be lower than the maximum output current rating of the charge controller. Most controllers for residential off-grid systems are rated for 60 to 80 amps.

Charge controller voltage IN: **90.9** volts (string operating voltage – from PV array)

Charge controller voltage OUT: **48** volts (battery-bank nominal voltage)

5. BATTERY BANK

Determine the number of batteries and the battery-bank wiring layout using the specs from your chosen model of battery, your daily electrical load, and your desired days of autonomy. You can quickly run these calculations to compare different battery types and models to determine the best fit for your application. The model shown here is an L-16-type, 6-volt, flooded lead-acid battery with a capacity rating of 390 amp-hours. FLA batteries should not be discharged more than 50%. Therefore, we will double the battery capacity to ensure a maximum depth of discharge (DOD) of 50%.

First, estimate the total battery storage capacity needed to meet the desired days of autonomy:

Days of autonomy: **3**

Average daily electrical load: **12.29** kWh

3 (DOA) × **12.29** (kWh/day) = **36.87** kWh

Next, double the desired storage capacity to factor for 50% DOD:

36.87 × **2** = **73.74** kWh

Convert to watt-hours by multiplying by 1,000:

73.74 × 1,000 = **73,740** watt-hours

Divide by the nominal battery-bank voltage to convert to amp-hours (watt-hours ÷ voltage = amp-hours):

73,740 watt-hours ÷ **48** volts = **1,536** amp-hours

Next, use the total amp-hour value to determine the total number of batteries and the battery bank layout. Pick any battery to run your initial calculations; then you can try different batteries to see how they affect the design.

Battery specs:

Model: **L-16**

Individual battery voltage: **6** volts

Battery capacity: **390** amp-hours

Divide the total amp-hour requirement by the battery's amp-hour rating to determine the number of parallel strings required (remember that each series-string of batteries has the same amp-hour rating as a single battery):

1,536 ÷ **390** = **3.94**; round up to **4** parallel series-strings

Each series-string needs eight 6-volt batteries to reach the nominal battery-bank voltage of 48 volts; therefore:

8 batteries per string × **4** parallel strings = **32** total batteries

Remember that all of the series-strings must have the same number of batteries (just as with modules in series), so your battery total must be divisible by 8. In this case, 32 is divisible by 8. The series-strings are wired together in parallel, so their amperage ratings add up. Calculate the total amp-hour capacity for the chosen battery model:

For 4 parallel strings of 8 batteries each:

390 amp-hour rating × **4** (strings in parallel) = **1,560** amp-hours

1,560 amp-hours exceeds your goal of 1,536 amp-hours, so this battery bank is sufficient to provide three days of autonomy (again, with a maximum DOD of 50%).

10 Beyond Installation:

Tips for Troubleshooting, Maintaining, and Monitoring Your PV System

GOAL

Learn how to shut down and start up your PV system, troubleshoot simple problems, and monitor everyday system performance

ONE OF BEST THINGS ABOUT ELECTRICAL SYSTEMS is that they rarely need maintenance. They don't spring leaks, need filter changes, get clogged, or beg for tune-ups. They don't fade or peel or rust. They don't even wear out in the foreseeable future. And for the most part, PV systems are no different. As far as regular maintenance goes, there's only one standard item on the to-do list: cleaning the modules every so often. If you feel like it. Yes, off-grid systems have batteries, some of which require maintenance, and all of which need replacement down the road (but you knew that going into this, and probably feel it's a reasonable price to pay for energy independence). This is not to say that PV systems never have problems. Of course things can go wrong, but when they do, the solution tends to be simple and usually involves replacing an interchangeable part. Then your system goes back to serving you – silently, reliably, without being asked. Much like the sun itself.

PV Safety Rules

Before you do any work on a fixture or outlet on your regular household AC electrical system (not your PV system), you switch off the breaker (or remove the fuse) on the circuit containing the fixture or outlet, and that reliably shuts off the power to the device. If you need to work on multiple circuits or you aren't sure which breaker to use, you can shut off the main breaker to shut down all of the circuits in the household system. However, the power from the utility service lines (and the terminals they connect to in the main electrical service panel) remain live unless the utility itself shuts down the electrical service to your house.

PV systems follow similar rules. You use disconnect switches to shut down most of the system, but the PV array remains live and is always producing electricity during daylight hours. Batteries in off-grid systems are always live unless they have been 100% discharged. This is why it's important to understand which parts are — and are not — isolated from the array and other power sources when you shut off the disconnects.

First, we'll take a look at PV safety rules; then we'll walk through the shutdown procedure for each type of PV system.

PV SAFETY RULE #1:
Always shut down the PV system and confirm the power is off before working on any wiring.

Even small residential PV systems produce more than enough electrical current to kill you. Accidents are totally preventable if you effectively shut off the power before working on or maintaining the system. Follow the proper procedure to shut down the entire PV system (see pages 173 to 175). If you will be working on or near any wiring, test the wires with a multimeter to confirm that the power is off before touching any wires.

PV SAFETY RULE #2:
Never connect or disconnect wires under load.

Load is a general term for anything that has the effect of drawing power from an electrical

Electrical arcing can give you a shock even if you have no direct contact with the electrical conductors.

circuit, creating a flow of current. It's dangerous to interrupt the circuit while the current is flowing.

If you've ever pulled a lamp cord out of an outlet while the lamp is still on and seen a spark at the end of the plug, you've witnessed electrical *arcing*. This is the flow of electricity literally jumping through the air between the outlet and the plug. The same thing can happen when you plug in a lamp that's already switched on. The electricity starts flowing before the plug is fully in place, creating an arc.

While disconnecting or connecting a lamp under load gives you a little spark, doing the same with a string of PV modules (especially a high-voltage string) can produce a really **big** spark, something you might describe as an electrical explosion in your face. Prevention of arcing hazards is simple: shut off the power to the PV system before messing around with any wires or connections. This makes PV Safety Rule #2 a lot like Rule #1, but it's important to understand how, thanks to arcing, you can be hurt by electricity even when you're touching only the insulated parts of wires, fittings, and so on.

Shutdown and Startup Procedures

The main shutdown devices for PV systems are the disconnect switches and the breakers. On most systems, the AC disconnect can be "locked out" in the OFF position by securing the switch arm in place with a loop of stiff wire or, better yet, a padlock or an electrician's lockout device. Locking out the AC disconnect switch prevents another person from turning on the power feeding the house while work is being done on the household system. If a switch has lockout capability, use it.

Different types of systems have somewhat different shutdown procedures, and it's import to understand what part of the system is disabled (at zero voltage) and what parts are not. For example, on a microinverter system, turning off the AC disconnect shuts off the power at the microinverters, which effectively shuts down everything but the solar modules themselves and their wire leads. With a string inverter or off-grid system, activating the rapid shutdown disconnect turns off the power at the disconnecting combiner box, which is within 10 feet of

the array. The array and all the wiring between the modules and the combiner box remain live. The only way to make a solar module stop producing electricity is to cover it completely so that no light gets through. But that is rarely necessary for normal system maintenance. Off-grid systems also have batteries, and batteries cannot be shut down; they contain dangerous amounts of energy unless they are 100 percent discharged.

When it's time to turn the PV system back on, it's important to follow the proper sequence. The standard procedures are covered here, but always follow the manufacturers' instructions for both shutdown and startup steps.

Electricians and other pros who work with electricity have a simple safety technique for turning switches on or off: stand to the side of the switch, looking away from it, and operate the switch with one hand only. This is a good habit to adopt.

HOW TO SHUT DOWN GRID-TIED SYSTEMS WITH STRING INVERTERS

1. Activate the shutdown switch or button on the system's rapid-shutdown control box to shut off the power between the array disconnecting combiner box and the inverter.

2. Switch the DC disconnect on the inverter to the OFF position (these switches typically do not have lockout capability).

3. Open the door to the main electrical service panel and switch the PV breaker(s) to the OFF position. Do not remove the inner, "dead front" panel inside the service panel box.

4. Turn off the DC–AC inverter. Some inverters have ON/OFF switches; others turn off when the AC disconnect is turned off (step 5).

5. Move the switch on the system AC disconnect to the OFF position, and lock it out.

WARNING: The rapid-shutdown control cuts all outgoing DC power at the disconnecting combiner box near the array. The array and the string home runs leading to the combiner box remain live and carry voltage during daylight hours. Because the system is grid-tied, there's still power in the main electrical service panel, the household electrical system, the net meter, and the utility's service entrance cable.

TO START THE SYSTEM:

1. Turn on the DC–AC inverter, as applicable.

2. Switch the AC disconnect to the ON position.

3. Turn on the PV breaker(s) in the main electrical service panel, and shut the panel door.

4. Manually reset the switch arm on the disconnecting combiner box at the array. You have to do this at the array box, not at the ground-level control box.

5. Turn the DC disconnect switch on the inverter to the ON position.

6. Wait for the inverter to boot up and display its main screen.

7. Check the inverter's readout to confirm the system is operating normally.

HOW TO SHUT DOWN GRID-TIED SYSTEMS WITH MICROINVERTERS

1. Open the door to the main electrical service panel and switch the PV breaker(s) to the OFF position. Do not remove the inner, "dead front" panel inside the service panel box.

2. Move the switch on the system AC disconnect to the OFF position, and lock it out.

WARNING: This shutdown procedure turns off the AC power at the microinverters, so there's no voltage in the system components between the microinverters and the PV breaker in the main service panel. Because the system is grid-tied, there's still power in the main electrical service panel, the household electrical system, the net meter, and the utility's service entrance cable.

TO START THE SYSTEM:

1. Turn on the AC disconnect.

2. Turn on the PV breaker(s) in the main electrical service panel, and shut the panel door.

3. Check the data monitoring system to confirm the system is operating normally.

HOW TO SHUT DOWN OFF-GRID SYSTEMS

1. If the system has a generator, turn off the generator disconnect to shut down AC power input and disable any auto-start function(s).

2. Activate the shutdown switch or button on the PV system's rapid shutdown control box. If the off-grid system does not have rapid shutdown (many do not), turn off all DC disconnects from the PV array.

3. Move the switch on the system AC disconnect to the OFF position, and lock it out. This ensures that the inverter is not under load. **Never shut down an inverter under load!**

4. Switch the DC disconnect *for the inverter* to the OFF position (these switches typically do not have lockout capability). This ensures the inverter cannot provide AC output.

5. Turn off the DC *main disconnect*, which will shut down the PV charge controller.

> **WARNING:** The rapid-shutdown control cuts all outgoing DC power — created by the modules, not the batteries — at the disconnecting combiner box near the array. The array and the string home runs leading to the combiner box remain live and carry voltage during daylight hours. The batteries and all wiring connected to the batteries remain fully energized.

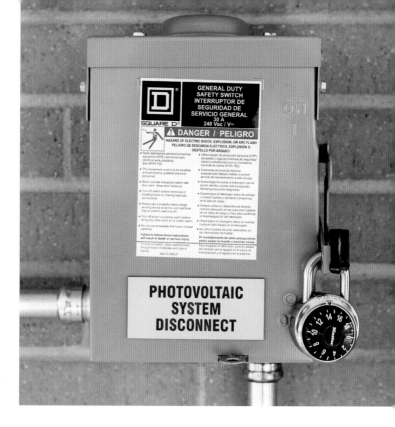

TO START THE SYSTEM:

1. Turn on the DC *main disconnect*, which will turn on the charge controller and/or battery-bank breakers.

2. Switch the DC disconnect *for the inverter* to the ON position.

3. Move the switch on the system AC disconnect to the ON position.

4. Manually reset the switch arm on the disconnecting combiner box at the array, if the system has rapid shutdown; you must do this at the array box, not at the ground-level control box. If the system does not have rapid shutdown, turn on all DC disconnects from the PV array.

5. If the system has a generator, turn the generator disconnect to ON and enable any auto-start function(s), as applicable.

> **WARNING:** The batteries are always the last to shut down and the first to connect at startup.

Maintenance and Troubleshooting

With no moving parts (unless you have a ground-mount with tracking), grid-tied PV systems need very little upkeep. Occasional cleaning, periodic visual inspection, and on-going monitoring of the array output are all that's required in terms of routine maintenance and system monitoring. If something seems amiss, there are a few things you can check before calling a solar pro or an electrician for help. Off-grid systems involve the same simple maintenance program for the array but also include maintenance duties for the batteries (flooded lead-acid batteries only).

Cleaning and Inspecting the Array

A routine cleaning not only helps optimize system performance but also allows for an up-close visual inspection of the modules, wiring, and other hardware, just to make sure everything looks okay. How often you need to clean depends on the local conditions. If the weather is dry and there's a lot of dust in the air, you may find that it's helpful to clean a few times a year. Heavy accumulation of tree pollen also warrants cleaning, as do bird droppings and leaves from trees, which effectively shade modules. But with many systems, if you decide to skip a seasonal cleaning or two, it probably will have little effect on your average system performance, especially if rain occasionally washes the modules for you. And, if necessary, you can always use binoculars to inspect a rooftop array from the ground.

Snow removal is also a good idea, but in general, once the sun comes out PV modules tend to lose their snow cover faster than conventional roofing. If you really want to sweep snow off of your modules, take safety precautions when getting onto a snow-covered roof.

1. Safety first! Shut down the system (see pages 174 to 175 for your system type). Use fall-arresting equipment for all rooftop work (see Rooftop Safety, page 98).

2. Clean the modules from the top down, using plain water and a nonabrasive sponge or coarse rag. You can use a synthetic scrubby pad for stubborn spots, but don't use metal scrubbies (or metal tools of any kind) or any abrasive cleaners, which can scratch the module glass.

3. Rinse the modules with plain water.

4. Squeegee the glass to remove water droplets. (Of course, you don't squeegee after every rainstorm, but rainwater has very little mineral content, unlike "hard" municipal water, which leaves mineral deposits.) You might want to use a long-handled window cleaning tool (with a sponge and squeegee on one head) to reach interior modules.

5. Inspect the modules and wiring for problems (see What Can Go Wrong with Modules and Wiring, opposite) and for evidence of small animals visiting or nesting under the array, potentially chewing on the wiring. If this is a problem, you can always critter-proof your array (see page 41).

Troubleshooting

The easiest way to check on the health of your PV system is through regular data monitoring, which allows you to keep tabs on performance factors like power output, array voltage and current, and kWh energy production. If you get into the habit of checking the data regularly (you don't have to geek out on it), you'll quickly discover patterns of normal function and learn how your system performs under different conditions. This becomes your basis for knowing when something *isn't* normal, such as when output is significantly down on a clear day. When problems arise, here's some easy stuff to check before calling a pro.

IF YOUR ARRAY PRODUCTION/POWER OUTPUT IS LOWER THAN NORMAL, IT COULD BE DUE TO ONE OR MORE OF THE FOLLOWING:

◇ Shading

◇ Dirty modules/debris on modules

◇ An unusually high ambient temperature (lowers module efficiency)

◇ Damaged wiring

◇ A bad module (may affect an entire string)

◇ Old batteries (for off-grid systems, batteries may need replacement every 5 to 8 years)

WHAT CAN GO WRONG WITH MODULES AND WIRING

Under normal conditions, a quality PV module placed online today will most likely be cranking out electricity at nearly the same rate 30 years from now. When problems do occur, it usually means replacing the module. Here are some of the problems you might see and what they can indicate:

- **CRACKED GLASS.** Cracks can let water inside the module, possibly leading to corrosion, electrical shorting, and even a shock risk; replace the module.

- **DISCOLORATION OF CELLS.** Cells with a cloudy, whitish appearance may indicate water intrusion, hot-spotting (at right), or delamination of the cell and glass, although delamination is much less common now than in the past.

- **CRACKED/BROKEN CELL.** Physical cell damage often is due to mishandling and is caught during installation, but things like hail and temperature stress can lead to cracked or broken cells after installation. The cell bus bars (the two metal ribbons or strips running vertically over the cells) often hold together broken cells so they continue to work. Otherwise, cell breakage can create an open circuit that puts the cell (not the whole module) out of commission.

- **HOT-SPOTTING.** Hot-spotting can occur when one cell in a string of cells is consistently shaded, leading to a concentration of power in a small area. Visual clues may include cell discoloration, corrosion or dark spots on cell wiring or bus bars, melted solder on intercell wiring, or cracked glass. Most modules today are designed with bypass diodes to help eliminate problems due to shading.

- **DAMAGED MODULE/STRING WIRING.** Chewed wiring is likely the work of squirrels, raccoons, or other varmints. Otherwise, damaged or disconnected wiring may be the result of sliding snow and ice, high winds, branches or other debris, or even abrasion against the roof surface, if the wiring is loose and touches the roofing. Replace any damaged wiring, and make sure it's secured better than its predecessor. Keep out animals by installing edge screening (see page 41).

IF THE SYSTEM IS NOT PRODUCING POWER AT ALL, IT COULD BE DUE TO ONE OR MORE OF THE FOLLOWING:

◇ A tripped breaker (visually inspect system for obvious problems, then reset breaker; if it trips again, call a pro)

◇ Blown fuse(s) in the combiner box

◇ Activation of the rapid shutdown (perform a restart to check)

◇ A failed control board in the string inverter

◇ A failed microinverter

◇ Someone (such as a contractor) shut off the PV system while working on the home's electrical system and didn't restart the PV system.

IF AN INDIVIDUAL MODULE HAS GONE OFFLINE (MICROINVERTER OR DC OPTIMIZER SYSTEMS ONLY), IT COULD BE ONE OR MORE OF THE FOLLOWING:

◇ A loose wiring connection or damaged wiring between a module and the microinverter or microinverter and the AC trunk cable

◇ A bad microinverter or optimizer (more likely)

◇ A bad module (less likely; experience has shown that while electronics and batteries fail, modules almost never do)

> **TIP**
>
> ## GET A GOOD VIEW
>
> Use a flashlight and a small mirror to look down into the battery cells to check water levels.

Battery Maintenance for Off-Grid Systems

Only flooded lead-acid batteries require regular maintenance (see note, below). Routine tasks include watering (refilling the batteries with distilled water), equalization charges, and testing for specific gravity and voltage. A quick visual inspection and cleaning of the battery terminals, as needed, can be added to any of the regular checks. Battery maintenance schedules and procedures are specific to each manufacturer and battery model; perform all maintenance as directed by the manufacturer. The following paragraphs offer a general description of what to expect.

NOTE: *While sealed batteries don't need regular maintenance, it's important to inspect them periodically to check for loose connections or corrosion and to clean the tops as needed. Follow the manufacturer's recommendations and inspection schedule.*

WATERING typically is required every 4 to 8 weeks, but schedules can vary widely. When batteries are new, it's a good idea to check the water levels once a month until you have a sense of how often they need topping off, and then you can adjust the schedule as needed. Refill only when batteries are fully charged, and use **only distilled water**. Follow the manufacturer's specifications for water levels.

If you check the water level when the batteries are not fully charged, and you find that the plates are exposed, add just enough distilled water to cover the plates. Charge the batteries fully, then check the water levels and add water to the level specified by the manufacturer.

WARNING: Batteries contain **lethal amounts of energy**. Be careful when cleaning the terminals and watering. Do not touch the terminals with your bare hands. Touching both terminals with a metal tool or other conductive material creates a short circuit that draws tremendous current capable of melting the tool and causing serious damage to anything in contact with it.

EQUALIZATION is a process of overcharging the batteries to reverse the effects of **stratification** (when acid levels become higher at the bottom of the battery than the top) and **sulfation** (a buildup of sulfate crystals on the plates), both of which shorten battery life. Follow the manufacturer's recommendations for how often you should equalize your batteries, which takes 2 to 4 hours. Most charge controllers can be programmed to initiate equalization on a set schedule. During equalization, check the specific gravity every hour until it no longer rises, at which point you should stop equalization. Because most PV arrays are not large enough to properly equalize batteries (and solar energy is weather-dependent), most people with off-grid systems run their generators for equalization.

SPECIFIC GRAVITY is the measure of the density of the electrolyte in the battery. A specific gravity test measures the battery's level of charge and can indicate problems that may result in incomplete charging. This is a simple test using an inexpensive tool called a hydrometer. Testing should be done periodically, following the manufacturer's recommendations. If the test indicates low specific gravity, the manufacturer probably will recommend that you perform an equalization charge.

safety goggles and rubber gloves (should be worn for all maintenance tasks)

distilled water and a paper cup or a glass with a spout (for watering batteries)

DISTILLED WATER

a mirror and a flashlight (optional)

baking soda and a wire brush (for cleaning terminals)

Baking Soda

hydrometer (for testing specific gravity)

MAINTENANCE AND TROUBLESHOOTING

179

Monitoring Your System

The old-school method of monitoring your PV system's production was to watch the utility net meter (hey, it's going backward!) and to look at the monthly electric bill (did we lower it and/or achieve net zero?!). But all this really tells you is how much solar energy you're producing relative to your energy consumption, on a monthly basis. If that's as much as you care to know, this monitoring method might suit you just fine. But be aware that if there's a problem that's affecting your energy production, you might not know about it until the next utility bill arrives.

The new-school method is to use a **data monitoring system**. This may sound complicated and geeky, but a monitoring system can be as simple as a smartphone app that reports your system's output in real time. That function alone can alert you to a problem with your PV system that may need attention. Here are some other monitoring tools available; some are included in smartphone apps, while others are functions of online software that you access with your computer:

- ◇ Present/latest power output (may be reported periodically, not necessarily in real time or instantaneously)

- ◇ Daily power output (Watts or kW versus time of day)

- ◇ Daily/monthly/total (historical) system production (kWh versus time)

- ◇ Array/string voltages and currents (helpful for identifying problems with individual module strings)

- ◇ Individual module performance (available only with microinverters and DC optimizers)

- ◇ Current and historic weather data

- ◇ Carbon emissions avoidance: how much CO_2 – and SO_2 and NO_x – you're keeping out of the air by using solar-generated power rather than nonrenewable energy sources

- ◇ Energy equivalents: comparing your solar production to other forms of energy; some software tallies how many trees you "planted" or how many cars you "removed from the road"

- ◇ Energy selfies that allow you to share your data online through social media (a real thing, except for the term *energy selfies* – we made that up)

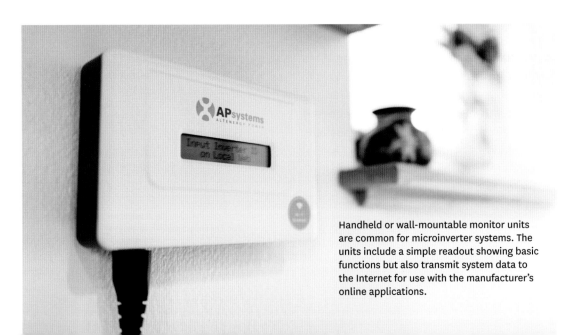

Handheld or wall-mountable monitor units are common for microinverter systems. The units include a simple readout showing basic functions but also transmit system data to the Internet for use with the manufacturer's online applications.

Data monitoring systems, services, and software are usually offered by inverter manufacturers. This makes sense because the inverter is the brain of the PV system. It knows how much DC power is coming in from the array and how much AC power is going out. Most of the other data points are just different ways of looking at this same information. Monitoring systems may be sold separately from the inverter(s), and may take the form of a handheld control unit or an add-on control board that's installed inside the inverter. You usually order this when you buy the inverter.

Connecting the unit or board to the Internet requires a Cat-5 cable (Ethernet cable) connected to your computer or wireless router, or you may be able to connect with a Wi-Fi device. Systems with multiple string inverters can be daisy-chained together with Cat-5 cable before connecting to a router.

Alternatively, if you have a string inverter, your monitoring system doesn't even need to be connected to the Internet; you can simply use the digital readout on the inverter unit to check your system status.

Of course, the best monitoring systems of all are your own eyes. If you need your sunglasses, you know you're making lots of electricity. May you continue generating electricity as long as the sun shines!

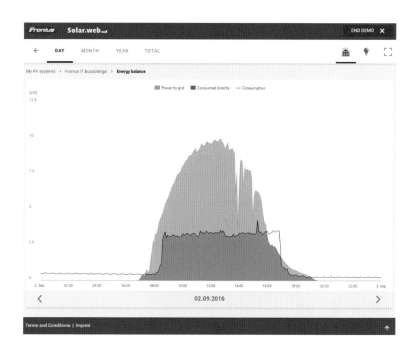

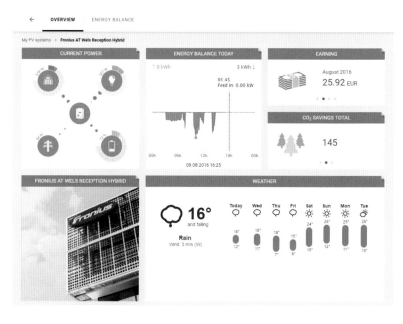

Data monitoring software programs for string-inverter systems report string currents and voltages, power output, and other critical information. In addition, they usually keep all the data since the system was initially started, so you can review your system's entire history.

ACKNOWLEDGMENTS

I would like to thank my parents, Beryl and Daniel Burdick, for a lifetime of love, support, and encouragement in all my endeavors, including my longtime pursuit of renewable energy solutions to the world's energy and environmental problems.

In addition, I further thank my siblings, Amy, Janet, David, and Lisa, as well as all my lifelong friends and colleagues, for their continual support and positive feedback as I followed my dreams, goals, and adventures.

I would like to thank all of my professional colleagues for their outstanding assistance with the preparation of this book, especially: Sam Anderson, Dan Fink, Tim Kjensrud, Tony Maday, Phil Myers, Tim Olsen, and Joe Thames. I would also like to thank all the people at Storey Publishing for their excellent work bringing this book to life. **— JB**

Thanks to Joe Burdick, Tony Maday, and the rest of the crew for showing me how it's done, from the drawing board to the first clean kilowatt. Thanks also to Hannah and Deb for their hard work on the book and to everyone at Storey Publishing for their old-school approach to a better future. And a well-deserved thanks to Troy Wanek for getting me started on the science and art of PV. **— PS**

For more information on PV system design and installation:

JOSEPH BURDICK
PV system design and installation
Burdick Technologies Unlimited, LLC
Lakewood, Colorado
burdicktechnologies.com

DAN FINK
Off-grid PV design and installation
Buckville Energy Consulting
Masonville, Colorado
buckville.com

TIM KJENSRUD
Master Electrician; PV and conventional electrical installation
Voltking Electric
Indian Hills, Colorado
voltkingelectric.com

TONY MADAY
PV installation and mechanical consulting
Red Rock Contracting
Parshall, Colorado
redrockcontracting.net

TIM OLSEN
Professional engineering services: solar, wind, and hydropower
Advanced Energy Systems, LLC
Denver, Colorado
windtechnology.com

GLOSSARY

AC disconnect Electrical switch in an outdoor-rated box that allows for manually disconnecting the AC electricity at a point between a string inverter and the main electrical service panel.

alternating current (AC) Electric current in which the direction of flow is reversed at frequent intervals (such as 50 or 60 cycles per second, known as 50 or 60 Hz). Most home electrical systems as well as most appliances and machines operate on AC electricity.

ampere, amp (A or I) Unit of electric current, the flow of electrons. One amp is 1 coulomb of electrons passing in one second. One amp is produced by an electric force of one volt acting across a resistance of one ohm. Abbreviated as A or indicated by the symbol I, or I(AC) (for AC current) or I(DC) (for DC current).

amp-hour (Ah) Quantity of electricity or measure of charge. One Ah means that one amp can flow or can be provided over a period of one hour. Battery capacity typically is rated in Ah.

array Group of PV modules connected together physically and electrically to provide a single electrical output. The solar-electricity-generating component of a PV system.

AWG American Wire Gauge, a standard system for designating the size of electrical wire. The higher the number, the smaller the wire diameter. Most household wiring is 12 AWG or 14 AWG; most PV circuits are connected with 10 AWG wire.

azimuth Compass angle, measured clockwise from north, that a solar array (or rooftop) faces: north is 0 degrees, east is 90 degrees, south is 180 degrees, west is 270 degrees.

battery bank Group of batteries, electrically connected together in series- and parallel-strings, to form a single electrical storage system.

charge controller Electronic device that regulates the voltage and current applied to the battery bank from the PV array. Essential for

ensuring maximum state of charge and longest battery life.

combiner box Outdoor-rated enclosure, usually located at or near the PV array, where PV module series-strings are electrically connected and combined into parallel strings, and where protection devices (such as fuses or breakers) are located. A *disconnecting combiner box* can be shut down remotely by a rapid-shutdown control device, shutting off the power between the combiner box and the DC–AC inverter.

critical loads panel Electrical service panel containing breakers for several "critical" circuits (such as those serving the refrigerator and other essential appliances). Only these loads will be powered by the backup system (batteries or generator) during a power outage.

crystalline silicon (c-Si or x-Si) Most common semiconductor material used in today's solar cells. It has a higher rate of efficiency (in converting sunlight to electricity) than thin-film photovoltaic materials and is produced in both monocrystalline and multi- (or poly-) crystalline forms. The first x-Si solar cell was made in 1954, by Pearson, Fuller, and Chapin, at Bell Laboratories in New Jersey.

current at maximum power (I_{mp} or I_{pmax}) The current at which maximum power (Pmax) is available from a photovoltaic cell or module.

cycle life Number of charge-discharge cycles that a battery can tolerate under specified conditions before it fails to meet specified performance criteria (e.g., battery capacity decreases to 80% of nominal). Represents the average lifetime of a battery.

days of autonomy (DOA) Number of days a house or building can operate normally on battery-backup power, if the sun doesn't shine (due to cloudy weather, etc.) and the PV system isn't able to recharge the batteries.

depth of discharge (DOD) Percent of discharge from a battery, ranging

from 0% (fully charged) to 100% (fully discharged). Most deep-cycle batteries used for PV systems should reach a DOD of no more than 50% under normal conditions. The current level of charge in a battery is often called the state of charge, or SOC.

direct current (DC) Electric current in which electrons flow in one direction only. Opposite of alternating current (AC). Photovoltaic cells and modules produce DC current when exposed to solar radiation from sunlight.

discharge rate The rate, expressed as amps over time, at which electrical current is taken from a battery.

electric circuit Path followed by electrons from a power source (such as the utility grid, a generator, battery, or PV array) through an external line (including any devices, appliances or machines that use the electricity) and returning through another line to the source.

electrical grid Integrated system of electricity distribution connecting utility power plants to substations, cities, and individual homes and other buildings. Also called utility grid.

electric loads Devices, appliances, machines, and so on, that draw power from an electrical system, such as the utility grid, PV array, generator, or batteries. Electrical wiring and other equipment is also considered a load because it promotes the flow of electricity through a circuit.

energy The ability to do work, or be stored for future use. Electrical energy is usually measured in kilowatt-hours (kWh), which is power (watts or kW) exerted or used over time (hours).

flush-mount Style of standard rooftop array installation in which PV modules are mounted to a module support structure (racking, in this case) so they are parallel to, and about 4 inches above, the roof surface.

footers Structural metal brackets that anchor rooftop racking to the roof structure. Typically installed over metal flashing (to prevent leaks) and fastened to the roof rafters or trusses with lag screws. They are purchased along with the racking system.

grid-tied PV system PV system that is connected to the utility grid and can both draw power from the grid and deliver it back to the grid. Also called grid-connected PV system.

ground-mount array Array of PV modules mounted on the ground instead of a rooftop. A ground-mount support structure typically includes vertical posts embedded in the ground and supporting an assembly of metal rails that hold the modules.

I_{mp} or I_{pmax} Specification symbol for current (amperage) at maximum power.

inverter Electrical device (usually in an outdoor-rated enclosure) that converts DC electricity into AC electricity. Different types of inverters are used for grid-tied and off-grid PV systems. A string inverter is a relatively large central unit that serves multiple PV modules or an entire array. Microinverters are small units that serve one or two modules each.

I_{sc} Specification symbol for short-circuit current.

kilowatt (kW) Unit of power equal to 1,000 watts, or 1,000 joules/second.

kilowatt-hour (kWh) Unit of energy equal to 1,000 watts acting over a period of 1 hour.

landscape orientation Positioning of PV modules in an array so that the long edges (lengths) of the modules run horizontally. Opposite of portrait orientation.

maximum power point (P_{max}) Point of highest possible power output of a PV module under given conditions, where the product of current (amperage) and voltage is maximum.

maximum power point tracking/tracker (MPPT) Power conditioning function or unit that automatically operates a PV module or array at its maximum power point (P_{max}) under all conditions (sunlight, temperature, and so on).

microinverter Small inverter that converts DC electricity to AC electricity for one or two PV modules. May be mounted onto module support structures or module frames. Also provides MPPT and monitoring at the module level.

module Solar industry term for a standard PV solar "panel." Typically describes a rigid-type unit with a metal frame and a clear glass cover protecting the layer of solar cells below. "Panel" may refer to units used with solar thermal (such as hot water) systems, while "module" typically indicates solar-electric or PV.

module mounting structure Hardware used to mount and secure the PV modules in an array. Applies to both rooftop and ground-mount arrays.

module specs Group of physical and electrical performance specifications for a PV module. Typically include maximum power (P_{max}), voltage at maximum power (V_{mp} or V_{pmax}), open-circuit voltage (V_{oc}), current at maximum power (I_{mp} or I_{pmax}), and short-circuit current (I_{sc}) as well as physical dimensions and number of solar cells. Also include certifications, such as UL1703, CEC, IEC1215, and so on.

NEC (National Electrical Code) Electrical code book published by the National Fire Protection Association (NFPA), typically revised every three years. Outlines recommended procedures, materials, and specifications for electrical systems in both residential and commercial applications. Essentially a rule book for ensuring proper design and installation of electrical equipment and systems. Most local code authorities (such as the local building department) adopt and enforce NEC specifications for all installations within their jurisdiction. The NEC includes specifications for most aspects of PV systems.

net meter *See* utility net meter.

off-grid PV system PV system that operates independently and is not attached to the utility grid. Solar-generated electricity is used to charge batteries, which in turn supply electricity to the home. Most off-grid systems include a fuel-powered generator for backup power when

the PV system cannot meet the household's electricity demand. Also called a stand-alone PV system.

ohm Unit of measure indicating resistance to the flow of an electric current. A 1-ohm resistor allows 1 volt of "electric force" to cause 1 amp of current to flow through the resistor.

one-line electrical diagram Document containing a simple diagram and specifications of all the electrical components in a PV system. Typically required to obtain a permit for PV system installation.

open-circuit voltage (V_{oc}) Maximum possible voltage across a PV module, produced when the module is in full sun while no current is flowing (amperage = 0). Measured at the unconnected positive (+) and negative (–) module leads while the module is not connected to an external load.

parallel wiring Wiring configuration joining together two or more PV modules, batteries, or module series-strings by connecting positive leads together and negative leads together. In the resulting circuit, the currents (amperages) from the units or series-strings add together, while the voltage of each unit or string in the circuit remains the same.

photon A particle of light (electromagnetic radiation) that acts as an individual unit of energy. Sometimes referred to as a "packet of light energy."

photovoltaic (PV) Pertaining to the direct conversion of light into DC electricity.

photovoltaic (PV) system Complete set of components of a solar-electricity-generating system that converts sunlight directly into electricity by the photovoltaic process. Includes the PV module array and the balance-of-system (BOS) components, including the module support structure, electrical wiring and grounding, and the DC and AC electrical components (inverters, breakers, disconnects, etc.) as well as the interconnection to the utility grid and/or batteries and generator, as applicable.

P_{max} Specification symbol for maximum power point. The point along a cell or module I-V curve where the I × V product (power) is the maximum.

pole-mount Type of ground-mount module support structure that is supported by a single (typically) vertical pole embedded into the ground. Includes top-of-pole and side-of-pole versions. Top-of-pole is standard for arrays that use tracking systems.

portrait orientation Positioning of PV modules in an array so that the long edges (lengths) of the modules run vertically. Opposite of landscape orientation.

racking Popular term for module support structure for rooftop or ground-mount PV arrays. Racking systems typically include footers (and flashing), rails, module mounting clips, and various bolts and screws.

rapid-shutdown system Safety system that allows utility workers, emergency responders, and other users to manually shut down a PV system, cutting power at a disconnecting combiner box (located near the PV array). When shutdown is activated (the circuit is open), there is no electricity in the PV system wiring or components between the combiner box and the main electrical service panel, but the PV modules and the wiring between the array and the combiner box remain live and can produce potentially lethal voltages during daylight hours.

self-islanding Safety feature found on all modern inverters that automatically shuts down the inverter if the utility power shuts down, such as during a power outage. Electricity remains live between the array and the inverter, but the (grid-tied) PV system cannot feed power to the utility grid. This is a safety requirement designed to protect utility workers and other personnel working on the power lines to restore electricity.

semiconductor Material with limited capacity for conducting an electric current. The level of conductivity generally falls between that of a metal and an insulator. Conducts electricity under certain conditions, such as upon exposure to voltage or light. Types of semiconductors suited for photovoltaic conversion include crystalline silicon, amorphous silicon, gallium arsenide, copper indium (gallium) diselenide, and cadmium telluride, among others.

series wiring/series-string Wiring configuration joining together PV modules or batteries by connecting the positive lead of one unit to the negative lead of the next unit in the series, creating a continuous string. In the resulting circuit, the voltages of the individual units add together, while the current (amperage) of the entire string remains the same as that of a single unit.

short-circuit current (I_{sc}) Current flowing freely from a PV module through an external circuit that has no load or resistance. Measured when the module's positive (+) and negative (-) leads are shorted together. The maximum current possible that can be produced by the module when exposed to full sun.

silicon (Si) Chemical element, atomic number 14, that is semi-metallic in nature. One of the most abundant elements on Earth and an excellent semiconductor material. The most common semiconductor material used in making solar cells.

solar cell The smallest unit/device capable of the photovoltaic effect: the act of producing DC electricity when exposed to light.

solar irradiance Sunlight, direct or diffuse, from incident solar radiation. Also a measure of solar radiation energy. The solar irradiance value used for STC testing is 1,000 Watts/ m^2 (1,000 watts per square meter). Sometimes called solar insolation (not to be confused with *insulation*).

split-phase Most common type of residential electrical service in North America; also called single-phase, three-wire. Supplies a total of 240 volts of power to a residence and can serve 120-volt devices on one "leg" of the supply source, or serve 240-volt devices on two legs. Commercial buildings in North America typically use 120/208-volt service; also called three-phase, four-wire. DC–AC inverters for PV systems often can be set to work with either system type.

STC (Standard Test Conditions) The conditions under which a PV module is typically tested in a laboratory. Basic values include (1) solar irradiance: 1,000 Watts/square meter; (2) cell (module) temperature: 25 degrees Celsius (77 degrees Fahrenheit); (3) solar reference spectrum: AM1.5 Global.

storage capacity The total energy stored in a battery or battery bank, usually specified as kWh (kilowatt-hours) of storage capacity.

string inverter *See* inverter.

sulfation Condition that afflicts unused and discharged batteries. Large crystals of lead sulfate grow on the plate(s), instead of the usual tiny crystals, making the battery extremely difficult to recharge.

tracking array PV array that follows the path of the sun to maximize production. Single-axis tracking rotates the array to track the sun from east to west, at a fixed tilt. A dual-axis (two-axis) tracking array points directly at the sun at all times by adjusting the east–west rotation and the array's tilt.

utility net meter Electric meter that moves both forward and backward to record electricity both used by, and produced by, the customer. The meter moves forward when the customer is using (buying) utility power and moves backward when the customer is selling solar-generated electricity back to the utility, thereby reducing the customer's electric utility bill.

voltage at maximum power (V_{mp} or V_{pmax}) The voltage at which maximum power (Pmax) is available from a photovoltaic cell or module.

V_{oc} *See* open-circuit voltage.

volt (V) Unit of measure for the force, or "push," given to the electrons in an electric circuit, causing current to flow through the circuit. One volt produces one ampere of current when acting on a resistance of one ohm. Abbreviated as V or indicated or V(AC) (for AC voltage) or V(DC) (for DC voltage).

watt (W) Unit of power, or amount of work done in a unit of time, measured in joules per second. In electricity, one ampere (A) of current flowing at a potential of one volt (V) produces one watt (W) of electric power; 1 amp × 1 volt = 1 watt.

watt-hour (Wh) Unit of energy equal to one watt acting over a period of one hour. *See also* kilowatt-hour (kWh).

185

METRIC CONVERSIONS

TO CONVERT	TO	MULTIPLY
inches	millimeters	inches by 25.4
inches	centimeters	inches by 2.54
inches	meters	inches by 0.0254
feet	meters	feet by 0.3048
feet	kilometers	feet by 0.0003048
yards	meters	yards by 0.9144
pounds	kilograms	pounds by 0.45

TO CONVERT FAHRENHEIT TO CELSIUS

subtract 32 from Fahrenheit temperature, multiply by 5, then divide by 9

INDEX

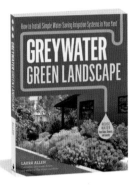

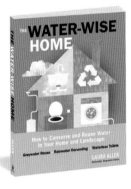